For Carmen and Rafael

What information consumes is rather obvious: it consumes the attention of its recipients. Hence, a wealth of information creates a poverty of attention and a need to allocate that attention efficiently among the overabundance of information sources that might consume it.

—Herbert Simon (1971)

Attention is a mode of payment, as well as the main input to scientific production. . . . Reputation is the asset into which the attention received from colleagues crystallizes.

—Georg Franck (1999)

Too often, the layman envisages scientific research as a locus of Olympian accord, an Arcadia of fairness. Teamwork in the sciences, in a patron's laboratory, can be fraught with jealousies, with fiercely competing egoism. Whose name will figure when the results are published? This *invidia* has become more acute as the economics of success become greater and the funding more precarious.

—George Steiner (2003)

Contents

Acknowledgments

Much, though by no means all, of the raw material for *The Hand of Science* derives from papers I have published or presentations I have given on scholarly communication and related topics in recent years. In some cases, the textual correspondence between the original and present versions is fairly easy to detect; in others, as a result of kneading, blending, and refreshing, the inheritance is no longer immediately obvious. For the record, I have listed the primary published sources below. A leitmotif of this book is the collaborative character of science and scholarship, whether the collaboration is formal or informal in nature. Truth be told, we rarely work as isolates; rather, we are embedded in a variety of social networks—invisible colleges, to use the time-honored term—and it is these webs of peer-to-peer connections that provide us with much of the stimulation and support essential for the development of our ideas and, ultimately, for the furtherance of our academic careers.

This is a periphrastic way of saying that there is a platoon of coauthors, trusted assessors, and backgrounded others, whose "beneficial collegiality," to use Laband and Tollison's (2000, 633) felicitous phrase, warrants acknowledgment. These good souls—*pace* Graham Harman (2002, vii) who feels that a long list of names at the front of a book is "like some Praetorian guard [that] often serves to intimidate readers, to make them feel outclassed by a competent network of college professors, research institutions, and fellowship foundations"—are, in alphabetical order, Helen Atkins, Katy Börner, Holly Crawford, Ron Day, Elisabeth Davenport, Martin Dillon, Rob Kling, Elin Jacob, Kathryn La

Barre, Geoffrey McKim, Stacy Nienhouse, Alice Robbin, Yvonne Rogers, Howard Rosenbaum, Debora Shaw, Lisa Spector, and Reyes Vila-Belda. My gratitude to them all, in appropriate measure. In addition, I should like to acknowledge the feedback I received from nameless colleagues following conference and seminar presentations at the ARL (Association of Research Libraries) in Washington, D.C.; Duke University; Indiana University, Bloomington; Manchester Metropolitan University; Napier University, Edinburgh; the University of New South Wales; OCLC (Online Computer Library Center), Columbus, Ohio; and the University of Toronto. Nothing *ex nihilo*.

REFERENCES

Harman, G. (2001). *Tool-Being: Heidegger and the Metaphysics of Objects*. Chicago: Open Court.

Laband, D. N. and Tollison, R. D. (2000). Intellectual collaboration. *Journal of Political Economy*, 108(3), 632–662.

ORIGINAL SOURCES

Cronin, B. (1992). Acknowledged but ignored: Credit where credit's due. *Bulletin of the American Society for Information Science*, 18(3), 25.

Cronin, B. (1996). Rates of return to citation. *Journal of Documentation*, 52(2), 188–197.

Cronin, B. (1998). Metatheorizing citation. *Scientometrics*, 43(1), 45–55.

Cronin, B. (1999). The Warholian moment and other proto-indicators of scholarly salience. *Journal of the American Society for Information Science*, 50(10), 953–955.

Cronin, B. (2000). Semiotics and evaluative bibliometrics. *Journal of Documentation*, 56(3), 440–453.

Cronin, B. (2001). Bibliometrics and beyond: some thoughts on web-based citation analysis. *Journal of Information Science*, 27(1), 1–7.

Cronin, B. (2002). Hyperauthorship: a postmodern perversion or evidence of a structural shift in scholarly communication practices? *Journal of the American Society for Information Science and Technology*, 52(7), 558–569.

Cronin, B. (2003). Scholarly communication and epistemic cultures. *New Review of Academic Librarianship*, 9, 1–24.

Cronin, B. (2004). Bowling alone together: Academic writing as distributed cognition. *Journal of the American Society for Information Science and Technology*, 55(6), 557–560.

Cronin, B. (2004). Normative shaping of scientific practice: The magic of Merton. *Scientometrics*, 60(1), 41–46.

Cronin, B. and La Barre, K. (2004). Mickey Mouse and Milton: Book publishing in the humanities. *Learned Publishing*, 17(2), 85–98.

Cronin, B. and Shaw, D. (2002). Banking (on) different forms of symbolic capital. *Journal of the American Society for Information Science and Technology*, 53(13), 1267–1270.

Cronin, B. Shaw, D., and La Barre, K. (2003). A cast of thousands. Co-authorship and sub-authorship collaboration in the twentieth century as manifested in the scholarly literature of psychology and philosophy. *Journal of the American Society for Information Science and Technology*, 54(9), 855–871.

Cronin, B., Shaw, D., and La Barre, K. (2004). Visible, less visible, and invisible work: Patterns of collaboration in twentieth century chemistry. *Journal of the American Society for Information Science and Technology*, 52(2), 160–168.

Chapter One

Scholars and Scripts

If I may paraphrase Monsieur Jourdain in Molière's *Le Bourgeois Gentilhomme*, for almost twenty-five years I have been engaged in a process of situated rhetorical action without really knowing it. This is what we do when we write, according to C. N. Candlin, general series editor of *Applied Linguistics and Language Study*, in his introduction to Ken Hyland's analysis of academic writing as "collective social practices" (Hyland 2000, 1). Just as in conversation, so it is with academic writing. Writing is a form of social interaction with our peers, and literary genres give shape and structure to those negotiated interactions. We don't usually insult colleagues to their face; nor do we usually seek to alienate them in our professional writings. Genre conventions, which are not immutable and can be subtly manipulated, are the disciplinarily sanctioned means whereby authors endeavor to convey their messages to readers. Academic writing is replete with genres and rhetorical tropes: this is a scholarly monograph (not easily confused with an abstract or a research article) and, as such, it should (and probably will) conform to the dominant structural, syntactic, and stylistic features of that genre, or at least one of its identifiable subgenres.

Hyland's corpus-based analysis of the various communicative categories, or moves, associated with different literary genres is instructive. Take the case of abstracts, where authors walk a fine line between providing the reader with an accurate representation of the larger text (often impersonal and seemingly agentless) and hooking the reader's interest using a certain degree of promotional legerdemain. In an

1

economy of attention, to use Simon's (1971) now very fashionable
phrase, the abstract has become an even more important instrument for
setting oneself apart from the mass of published work competing for the
reader's attention. Equally interesting is Hyland's analysis of a relative
newcomer in the stable of academic writing genres, the scientific letter
(e.g., *Physics Letters B*). Unlike the traditional, peer-reviewed article,
which eventually appears in the discipline's journal of record, the scientific letter (a return to the *Ur*-form of communication, as we shall see
in chapter 3) is a fast track for channeling breaking news to the scientific community. Here, boldness and tentativeness coexist. The author's
claims must be sufficiently compelling to hold the reader's attention,
yet not so brash as to constitute a breach of scientific reporting conventions. Hyland's (2000, 87) microlevel lexical analysis of this genre illustrates the role played by hedges and boosters ("communicative
strategies for increasing or reducing the force of statements") in main-
taining stability between these at times conflicting objectives. For example, scientific letters make much greater use of boosters (e.g., evi-
dently, clearly, obviously) than do conventional research articles, but
hedges (e.g., may, seem, possibly) are used more frequently in letters
than boosters.

Citation analysis, especially evaluative bibliometrics, is one of the
staples of information science research. Linguists, however, are typically more interested in the language forms associated with citations
than with their potential as tokens for use in research evaluation or science mapping exercises. Hyland provides a cross-disciplinary analysis
of the reporting verbs associated with integral and nonintegral citations.
He systematically analyzes how authors embed their arguments in networks of references (as I shall sedulously demonstrate in the pages that
follow) and how those referencing acts are linguistically framed in the
text. One is struck, however, by the differences between Hyland's approach and the approaches traditionally favored by information scientists. It is as if the two research communities are at times unaware of
each other's existence, an observation made originally by Swales
(1986). For example, and without any implied criticism, Hyland doesn't
mention Small's (1978) work on citations as concept markers/symbols
or White's (e.g., 2001) studies of citation image and identity (antici-
pated, as it happens, in Hyland's [2000, 37] notion of a "professional
persona"). Such omissions merely serve to underscore the desirability

of connecting these two largely noninteracting literature sets. As a rule, information scientists have tended not to exploit corpora or the range of applied linguistic research methods described by Hyland in his book. However, it is to be hoped that the gap between sociolinguistics and information science can be bridged; both sides have much to gain by coming together and trading insights, as demonstrated by White's (2004) bibliometric analysis of the interaction between the citation analysis and discourse analysis communities. *Disciplinary Discourses* is a useful addition to the burgeoning literature on academic writing, notably the foundational contributions of Bazerman (1988) and Swales (1990, 2004). By examining texts as disciplinary practices, Hyland shows how disciplines, from biology to marketing, are (re)produced by writing, and how different disciplinary practices and values are revealed in scholars' deployment of particular genre conventions.

But genre is only one facet of academic writing that is shaped by, and in turn shapes, disciplinary norms. Disciplines exhibit at times quite different sociocognitive structures, as we shall discover in chapter 2 ("Epistemic Cultures"), manifested in differing material practices, communication behaviors, and publishing regimes. By way of illustration, the global community of high-energy physicists is remarkably self-aware and essentially self-regulating (Galison 2003; Knorr Cetina 1999), which appears to render it immune to the kinds of publication abuses associated with some other scientific fields (see chapter 3 on the phenomenon of hyperauthorship and, specifically, problems of fraud in biomedical research). Collaboration and coauthorship may be routine in biomedicine, but they remain relatively uncommon in, for instance, philosophy or comparative literature. To take another example, the rules and institutional expectations of learned societies and professional associations relating to the posting, publication, and self-archiving of research vary greatly, even though the very same toolsets and platforms for authoring, distributing, and depositing original work are universally available. In short, scholars appropriate and co-opt information and communication technologies (ICTs) in myriad, often unpredictable, ways, reflecting underlying differences in their epistemic cultures (Biagioli and Galison 2003; Knorr Cetina 1999).

Technology does not determine scientists' modes of communication and publication, as sociotechnical systems analysts have long understood. Rather, the uses to which technologies are put are as much a function of the social values and choices of specific scholarly communities as of the functionalities and affordances of the toolsets themselves (Kling 1999). For historians, the interpretive monograph remains the dominant form of publication—and very much the gold standard as far as tenure is concerned (Estabrook 2003). This is certainly not the case in, say, economics (where refereed journal articles hold sway), physics (electronic pre-prints), or computer science (conference proceedings). These differences in working practices and tolerances also find expression in the ways disciplines choose to define the responsibilities associated with authorship—specifically, how they allocate credit and reward originality (Biagioli 2003). When it comes to tenure in literature and language departments, as we shall see in chapter 7 ("Symbolic Capitalism"), faculty members will be expected to have published a sole-authored monograph, whereas a number of rigorously peer-reviewed journal articles along with a record of continuous extramural funding will be the norm in domains such as astronomy or experimental psychology. Disciplinary differences shape not only the knowledge production but also the knowledge consumption practices of scientists and scholars. More specifically, the importance attached to citations as proxies of quality varies across disciplines, reflecting the ways scholars in different fields marshal evidence, construct arguments, and formally cite the work of their peers (see Najman and Hewitt [2003] on how disciplinary differences in information exchange and dissemination have a bearing on publishing and citation patterns).

The rate of change and experimentation within the world of scholarly communication during the last decade has been nothing less than remarkable, propelled by advances in web- and Internet-based publishing, the rapid growth in electronic journals, the open access (OA) movement, and the burgeoning idea of a digital scholarly commons. John Unsworth has captured the *zeitgeist* rather well:

> If the nineties were the e-decade (e-commerce, e-business, e-publishing, eBay, E*Trade, etc.), the aughties are the o-decade (open source, open systems, open standards, open archives, open everything). This trend, now unfolding with special force in higher education, reasserts an ideology, a meme, that has a continuous tradition traceable all the way back to the beginning of networked computing. . . . Call this meme Liberation Technology. (2004, B16)

Any one of these change drivers would be material enough for a book, which is a prelude to saying that the present volume—a short, synthetic, and copiously referenced *vade mecum*—offers neither a comprehensive account of academic publishing initiatives nor a definitive analysis of the structural dynamics of the scholarly communication marketplace. Rather, I revisit and refine ideas on the sociology of scientific writing developed in two earlier books, *The Citation Process* (Cronin 1984) and *The Scholar's Courtesy* (Cronin 1995), in the light of current trends in digital communication and academic publication. More particularly, I (i) examine the relationship between epistemic cultures and scholars' publication practices, (ii) consider the shifting nature of authorship, (iii) explore the significance of subauthorship collaboration, and (iv) assess the growing importance of symbolic capital in academic life. In the process, I draw upon a range of disciplinary literatures (communication studies, economics, information science, semiotics, and social studies of science to name a few) and raise a miscellany of questions, directly or indirectly: What does it mean to be an author in an age of collective effort? How are responsibility and credit allocated in collaborative endeavors? What is the relationship between reading, referencing, and reputation—the political economy of citation? How are social relations inscribed in intellectual space? Will the move to online and open-access publishing provide new measures of authorial salience and intellectual impact? My goal here is to try to capture the complex relationship between authorship (itself an increasingly problematic construct, as will soon become clear) and the reward system of science, understood in terms of an economy of attention. As Franck (1999, 54–55) argued persuasively in a *Science* paper a few years ago,

> Attention is a mode of payment, as well as the main input to scientific production. . . . Money is not the main motive for engaging in science. . . . Success in science is rewarded with attention. . . . Earning this "attention" income is a prime motive for becoming a scientist and practicing science. . . . The license for using somebody else's information productively is obtained through citation: in essence a fee paid through transfer of some of the attention earned by the citing author to the cited author.

I have written *The Hand of Science* in the hope that it will be read (an attention-demanding activity) and, ultimately, cited (a public affirmation of attention having been allocated) by my peers. I naturally want the book to be reviewed, and would much rather it received a panning than be ignored altogether. Sales figures and royalty payments per se are of relatively little import; direct pecuniary gain is assuredly not the motivation for putting pen to paper in this case. Rather, the primary goal—if I may mix my metaphors—is to attract eyeballs and harvest attention; in other words, to increase my stock of symbolic capital (Bourdieu l991). This kind of thinking, some feel, has led to "the cult of individual assessment in the university and the emphasis in academia on stockpiling refereed articles, even if hardly anyone reads many of them" (Guterson 2003, 301). Academic writing may indeed be a form of vanity fair, to draw on the title of Franck's (1999) aforementioned article, but self-interest and the commons are not inherently incompatible. Competition and communism, as Merton (1957, 659) long ago recognized, can normatively coexist: "When the institution [of science] operates effectively, the augmenting of knowledge and the augmenting of personal fame go hand in hand; the institutional goal and the personal rewards are tied together." Social constructivists may decry the lack of attention to individual practice, materiality, grounded observation, and situated action in sweeping, functionalist interpretations of science, but for some of us Mertonianism continues to offer a suave theoretical framework that melds institutional with individual motivations: "It will have been noticed that the instrumental and symbolic signaling functions of citations are not simply a matter of individual motivation but depend upon the historically evolving character of science as a social institution. The very form of the scientific article as it has evolved over the last three centuries normatively requires authors to acknowledge on whose shoulders they stand" (Merton 2000, 439).

Over the years I have quoted liberally from Merton's oeuvre, drawing on his normative framework (disinterestedness, universalism, etc.) to account for the demonstrable consistency of citation practices within the primary communication system of science (see Cronin 2004). Even if the rules of citation are not codified, scientists are presumed to have a tacit understanding of what is expected of them, just as seems to be the case with other forms of acknowledgment. In *The Scholar's Courtesy* (Cronin 1995, 107), I showed that "scholars subscribe to the idea of a governing etiquette" and exhibit "a highly socialized approach to acknowledgement." Merton would, I imagine, have put it somewhat differently, and certainly more elegantly. Indeed, in *The Normative Structure of Science* (Merton 1973, 276), he asserts that there exists "a distinctive pattern of institutional control of a wide range of motives which characterizes the behavior of scientists" and goes on to say that "once the institution of science enjoins disinterested activity, it is to the interest of scientists to conform on pain of sanctions and, insofar as the norm has been internalized, on pain of psychological conflict." I am inclined to think that Merton would have welcomed the communicative transparency of open-access publishing, in particular, and the opportunities it creates for developing novel forms of sociometric analysis and attention tracking, a view apparently shared by Boyle (2004): he entitled a recent paper on the emerging scholarly commons "Mertonianism unbound?"

"There are," as Montgomery (2003, 1) parsimoniously observes, "no boundaries, no walls, between the doing of science and the communication of it; communicating *is* the doing of science." New genres of academic writing and forms of authorship will continue to emerge, coevolving with both the material practices of scientists and advances in information and communication technologies. The ways and means by, which scientists and scholars create new knowledge, publish their ideas, and subject them to the scrutiny of their peers are undergoing continuous and occasionally significant change. As we shall see, behavioral diversity and heterogeneity of norms rather than standardization and uniformity are defining features of the contemporary scholarly communication and publication system. In short, disciplines are different, and those differences matter. One thing does, however, remain constant: the competition for attention—the ever-strong currency of academia. It is no exaggeration to say to say that the exhortation, "Publish or perish," narrowly construed, has become something of an anachronism in an age where "hits," downloads, "reads," citations and impact factors have become the default measures of a man's worth.

REFERENCES

Bazerman, C. (1988). *Shaping Written Knowledge: The Genre and Activity of the Experimental Article in Science*. Madison: University of Wisconsin Press.

Biagioli, M. (2003). Rights or rewards? Changing frameworks of scientific authorship. In: Biagioli, M. and Galison, P. (Eds.). *Scientific Authorship: Credit and Intellectual Property in Science*. New York: Routledge, 253–279.

Biagioli, M. and Galison, P. (Eds.). (2003). *Scientific Authorship: Credit and Intellectual Property in Science*. New York: Routledge.

Bourdieu, P. (1991). *Language and Symbolic Power*. Cambridge, MA: Harvard University Press.

Boyle, J. (2004). Mertonianism unbound? Imagining free, decentralised access to most cultural and scientific material. Paper presented at the *Workshop on Scholarly Communication as a Commons, March 31–April 2, 2004*. Workshop in Political Theory and Policy Analysis, Indiana University, Bloomington, Indiana.

Cronin, B. (1984). *The Citation Process: The Role and Significance of Citations in Scientific Communication*. London: Taylor Graham.

Cronin, B. (1995). *The Scholar's Courtesy: The Role of Acknowledgement in the Primary Communication Process*. London: Taylor Graham.

Cronin, B. (2004). Normative shaping of scientific practice: The magic of Merton. *Scientometrics*, 60(1), 41–46.

Estabrook, L. (2003). *The Book as the Gold Standard for Tenure and Promotion in Humanistic Disciplines*. Available at http://lrc.lis.uiuc.edu/reports/cic/CICBook.html.

Franck, G. (1999, October 1). Scientific communication—A vanity fair? *Science, 286*, 53–55.

Galison, P. (2003). The collective author: In: Biagioli, M. and Galison, P. (Eds.), *Scientific Authorship: Credit and Intellectual Property in Science*. New York: Routledge, 325–355.

Guterson, P. (2003). The death of the authors of death: Prestige and creativity among nuclear weapons scientists. In: Biagioli, M. and Galison, P. (Eds.), *Scientific Authorship: Credit and Intellectual Property in Science*. New York: Routledge, 281–307.

Hyland, K. (2000). *Disciplinary Discourses: Social Interactions in Academic Writing*. London: Longman.

Kling, R. (1999). What is social informatics and why does it matter? *D-Lib Magazine*, 5(1). Available at www.dlib.org/dlib/january99/kling/01kling.html.

Knorr Cetina, K. (1999). *Epistemic Cultures: How the Sciences Make Knowledge Work*. Cambridge, MA: Harvard University Press.

Merton, R. K. (1957). Priorities in scientific discovery: A chapter in the sociology of science. *American Sociological Review*, 22(6), 635–659.

Merton, R. K. (1973). The normative structure of science. In *The Sociology of Science: Theoretical and Empirical Investigations*, edited and introducted by N. W. Storer. Chicago: University of Chicago Press, 267–278.

Merton, R. K. (2000). On the Garfield input to the sociology of science: A retrospective collage. In: Cronin, B. and Atkins, H. B. (2000) (Eds.). *The Web of Knowledge: A Festschrift in Honor of Eugene Garfield*. Medford, NJ: Information Today, 435–448.

Montgomery, S. L. (2003). *The Chicago Guide to Communication Science*. Chicago: Chicago University Press.

Najman, J. M. and Hewitt, B. (2003). The validity of publication and citation counts for sociology and other selected disciplines. *Journal of Sociology*, 39(1), 62–80.

Simon, H. (1971). Designing organizations for an information-rich world. In: Greenberger, M. (Ed.). *Computers, Communication and the Public Interest*. Baltimore, MD: Johns Hopkins University Press, 37–72.

Small, H. (1978). Cited documents as concept symbols. *Social Studies of Science*, 8, 237–340.

Swales, J. (1986). Citation analysis and discourse analysis. *Applied Linguistics*, 7, 39–56.

Swales, J. (1990). *Genre Analysis: English in Academic and Research Settings*. Cambridge: Cambridge University Press.

Swales, J. (2004). *Research Genres: Explorations and Applications*. Cambridge: Cambridge University Press.

Unsworth, J. M. (2004, January 30). The next wave: Liberation technology. *Chronicle of Higher Education*, B16–B20.

White, H. D. (2001). Authors as citers over time. *Journal of the American Society for Information Science and Technology*, 52(2), 87–108.

White, H. D. (2004). Citation analysis and discourse analysis revisited. *Applied Linguistics*, 25(1), 89–116.

Chapter Two

Epistemic Cultures

The ways in and means by which scientists communicate have long attracted scholars' attention (Meadows 1998). Structural-functionalist accounts of how science works as a social system (Merton 1976) coexist (and sometimes compete) with thickly descriptive laboratory accounts of science-in-action (Latour and Woolgar 1979), discipline-specific mappings of communicative processes (Garvey and Griffith 1971), and painstaking sociohistorical accounts of scientific practice (Shapin 1994). Science, often viewed as "the premier knowledge institution throughout the world" (Knorr Cetina 1999, 1), has been studied intensively, and scientists themselves are a much-observed species, both from afar and also up close in their natural habitats. Knowing how scientists work, how they interact with their peers and publics, is not just intrinsically interesting to ethnographers, sociologists of science, and sundry others but has a bearing on the development of effective academic information resources and information support systems. Why that is the case will shortly become clear.

A great deal of scientific communication—indeed, scholarly communication in general—is informal in nature. Information diffusion depends on the conductivity of sociocognitive networks, sometimes referred to as "invisible colleges," a term first used in the seventeenth century (Crane 1972; Cronin 1982). Nevertheless, the ultimate goal of science, to appropriate the language of Latour and Woolgar (1979), is to produce inscriptions—to publish. Scientists, of course, are motivated by curiosity, intellectual excitement, and discovery, but publication is the means whereby the results of their cogitations and endeavors move into the public sphere. The formal publication practices and preferences of scientists are also well documented (Meadows 1974; Friedlander and Bessette 2003). Indeed, the quantitative analysis of the dynamics of scholarly publishing, most commonly and traditionally known as bibliometrics, has broadened in scope to become an academic subfield (scientometrics being the preferred descriptor) in its own right with a robust research literature (think, for instance, of the eponymous journal *Scientometrics* and its web-based cousin *Cybermetrics*). But the work of scientometric pioneers such as Derek de Solla Price (Price 1961, 1965) is of interest to more than a select band of researchers in information science and cognate domains. Moreover, there is an applied dimension: bibliometric tools and techniques can be used to inform collection development and collection management policies in academic libraries. More generally, bibliometrics provides insights into the communicative dynamics of science and, thus, can help librarians better understand the evolution, structural characteristics, and interactions of the disciplinary corpuses they manage, which, presumably, translates into more responsive services for academic end-users. With the advent of the web, classical and novel bibliometric measures—webometrics, netometrics, cybermetrics, and sitations are just some of the associated neologisms—are being developed and used to quantify online communication phenomena and behaviors (Cronin 2001a; Thelwall, Vaughan, and Björneborn 2005), a topic we consider at greater length in chapter 8 ("The Attention Economy").

VOLATILE ECOSYSTEM

But you don't have to be a born-again bibliometrician to generate useful trend data. The ARL (Association of Research Libraries) has produced a set of indicators on monograph and serial costs in North American universities that has achieved near iconic status both within and well beyond the academic library community (see www.arl.org/stats/ arlstats/graphs/2002/20002t2.html). These statistics and related extrapolations show a seemingly ineluctable trend dating from the mid-1980s: journal prices (in science, technology, and medicine, especially) are outpacing the CPI (Consumer Price Index) to such a degree that library budgets cannot absorb the strain (see also Cummings et al. [1992] for an earlier assessment of the phenomenon). Between 1986 and 2002, serial unit costs rose by 227 percent, while the number of serials purchased by academic libraries grew a mere 9 percent. During the same period, monograph expenditures increased 62 percent, while the number of books acquired by libraries decreased 5 percent. The ARL has also produced a set of supplementary statistics covering electronic resources. The most recent data show that expenditures for electronic resources account for 26 percent of ARL library materials budgets (see www.arl.org/newsltr/235/snapshot.html). Expenditures for electronic serials have increased by 75 percent in the past two years, and almost nine-fold since the mid-1990s. But one should remember that similar concerns have been voiced before. White's (1977) survey of over 400 libraries showed that serials subscription costs rose on average 13 percent per annum for the years 1969–1973.

The ARL statistics provide powerful ammunition for those who maintain that there is a "crisis in scholarly publishing" and that "publishers have been able to get away with economic murder" (Guédon 2003, 130). Judging by the frequency with which the word "crisis" appears in the titles of conference papers, journal articles, and technical reports, and also the regularity with which the ARL data are cited by librarians and proponents of the open-access movement (for instance, Harnad's July 13, 2003 posting to the SEPTEMBER98-FORUM speaks of "a genuine serials pricing crisis"), this is anything but a minority view. Dissenting voices, it seems, are relatively rare, but they do nevertheless exist.

According to Ayris (n.d.), "Sir Brian Follett says the escalating costs in periodicals and subscriptions are marginal to a university." In fact, as I discovered, Sir Brian (chairman of the Follett Report and Implementation Groups on Information Technology in the U.K.) said something slightly different. His actual words were, "I'm not optimistic about influencing journal prices. They are a marginal cost to research" (see www.sconul.ac.uk/event_conf/interconsortia2002/summary.html). Presumably, if they are marginal to the cost of research, they are even more marginal to a university's general operating fund. (Scientists and humanists would react differently to the use of the word "marginal" in this context, I imagine.) A quick back-of-the-envelope calculation illustrates the essential point. Indiana University's Bloomington campus had an expenditure budget of $900 million for FY02–03. The university library's slice of the pie was $30 million, of which $10 million was allocated to materials. In all, $5 million were spent on serials, roughly 0.5 percent of total campus outlay. Understandably, university presidents and vice chancellors may not be inclined to invoke the rhetoric of organizational crisis for oscillations in what constitutes a relatively modest line item in the institution's base budget.

Whether the ARL data describe a credible crisis, or whether it is the case that librarians are parochially oversensitized, can be debated ad nauseam. The fact of the matter is that the ARL numbers do not tell the whole story; rather, they are a surface manifestation of deep changes taking place in the scholarly communication ecosystem. Nor has this development occurred overnight. Many of us have had a vague sense that the Old World order was set to change, without necessarily being clear as to the contours of the emerging New World order. Fifteen years ago I wrote as follows (Cronin 1989, 24):

> Academic researchers give the fruits of their intellectual labour (conveniently packaged in article form) to publishers who sell them back (bound between covers) to the same population of individual scholars/authors. A bizarre system, in which the resultant profits are in no way commensurate with the value added, or the business risk taken, by the middlemen. This historical dependency can be broken, but it will require a sustained and unusually high degree of co-operation between authors, universities and learned/professional bodies. The solution is forward integration, with learned societies and/or universities becoming mainstream publishers. . . . An alternative would be the foundation of a number of university publishing consortia, or joint ventures involving universities and learned societies.

At that time, the lexicon of open access, self-archiving, and institutional repositories hadn't been created. By the late eighties and early nineties, pathbreaking initiatives such as SPARC (Scholarly Publishing and Academic Resources Coalition, see www.arl.org/sparc/home/index.asp?page=0) were still not even a twinkle in the library profession's eye, but we were at least taking the first tentative conceptual steps toward the kinds of radical, collectivist solutions (e.g., BioMed Central, Public Library of Science [PLoS]) that today seem set to shake up, if not in fact alter irrevocably, stakeholder relationships in the scholarly publishing marketplace. Most revealing, though, was my use of the reductionist phrase "[t]he solution is forward integration." There is *a* crisis (escalating periodical prices) for which there is but *one* solution (vertical integration). I seemed to feel then that there was a single model of scholarly publishing and if that model was broken, it was a straightforward matter of slotting another into its place. With hindsight, this kind of one-size-fits-all thinking seems wholly inappropriate given the multidimensionality of the wider scholarly communication system. A decade or so on and I am espousing a more nuanced view (Cronin 1999a, A25):

> Disciplines differ significantly in terms of their sociocognitive structures, degrees of paradigmatic consensus, funding mechanisms, collaborative intensity, and institutionalized quality assurance mechanisms. . . . Given the diversity of stakeholders and heterogeneity of culturally validated practices, it would be foolish to imagine that a monolithic e-publishing model or system could emerge. Pluralism, plasticity, and adaptivity will be hallmarks of the new world order.

How one's perspective changes over time.

TRIBAL CUSTOMS

A number of factors has contributed to the sometimes simplistic conceptions of scholarly communication and publishing that prevail. First, the communication ecosystem is highly diverse, consisting as it does of many species—academic tribes, to use Becher's (1989) term—characterized by often markedly different behaviors and drawing upon a bewildering array of (information) resources and, in turn, a rich array of institutional arrangements for the management of those resources.

Second, the velocity and variety of experimentation in electronic and online publishing (think of TULIP, Project Muse, PEAK, HighWire Press, ELectronic Society for Social Scientists, BioMed Central, and numerous other more or less persistent and influential initiatives) can be overwhelming. There is continuous experimentation and many plausible development trajectories suggest themselves: in short, a patchwork of possibilities—of electronic publishing futures—exists (Wellcome

Trust 2003). Innovations may be ten a penny, but consolidation and co-
herence are still notable by their absence.

Third, there is a lack of discursive consistency and semantic stability.
What exactly does it mean to publish in the digital age? What is an electronic journal? What do we mean by the term "eprint"? For some, these
are unproblematic constructs, but Kling and colleagues (e.g., Kling and
Callahan 2002, 134) have demonstrated that there is a degree of sloppi-
ness in our use of many of the key terms routinely deployed in discussions of electronic publishing and scholarly communication. By way of
illustration, they distinguish between four kinds of electronic journals:
pure e-journals (distributed only in electronic form), E-p-journals (primarily distributed in electronic form but with limited distribution in paper form); P-e-journals (primarily distributed in paper form); and P+e
journals (initiated with parallel paper and electronic editions). This is
not an exercise in pedantry, but an attempt to unbundle the idea of the
electronic journal and to better understand the interactions between
publishing technologies on the one hand and the behaviors and value
choices of scholars/users on the other. In his 2004 *ARIST* (*Annual Review of Information Science and Technology*) chapter, Kling (2004) dis-
ambiguates foundational terms such as "manuscript," "publication,"
"preprint," "article" and "working paper"; analyzes a variety of archi-
tectures used for communicating "unrefereed e-scripts"; and considers
why such practices are, for now at least, confined to a minority of academic disciplines. Kling (Kling and Swygart-Hobaugh 2002) has also
examined the effect of the Internet on the velocity of scholarly journal
publishing, concluding thus:

> "[O]ur results challenge the belief that the Internet has speed [*sic*] up
> scholarly communication through journals across all disciplines. While the
> Internet may have possibilities for increasing the efficiency of the publica-
> tion process, several other factors, such as publication volume, differential
> acceptance and/or exploitation of Internet capabilities, disciplinary differ-
> ences in demand for rapid publication, peer-review processes, submission
> rates, etc., also play a significant role in this process."

In these and related studies (e.g., Kling and McKim 2002), Kling has
demonstrated how careful empirical analysis can expose fallacies in
thinking related to scholars' electronic publishing and digital communication practices.

Fourth, there is a tendency to talk deterministically with regard to the effects and impacts of ICTs (information and communication technologies) on scholarly publication practices, as if options and outcomes were shaped exclusively by the features, affordances, and functionalities of available tool sets, or, alternatively, to assume that the aggregate institutional apparatus of science and scholarship exhibits a path-dependent pattern of development (see North [1990] on institutional change and path dependency). Kling and McKim (2002, 6) refer to this general tendency as "information processing theorizing," an approach that "elides and homogenizes field differences." Although other scholars have applied sociotechnical systems approaches to developments in ICTs (Bishop and Star 1996), Kling has done much to sharpen our intuitive sense of how the technologies of electronic publishing and disciplinary publishing regimes are coconstitutive.

A fifth factor that encourages oversimplified understandings of electronic publication is the progressive privatization of higher education, which has been accompanied by the unseemly expansion of "the bailiwick of the apparatchiks" (Cronin 2001b, 133). In the United Kingdom, the new managerialism has its own lexicon: the world of learning is now referred to without irony as "the Sector." The primary allegiance of career administrators is to corporate accounting, not disciplines. In such an environment, it is not hard to see why across-the-board/off-the-shelf/one-size-fits-all solutions are likely to find favor.

How do scientists create new knowledge, and why should we care? Knorr Cetina (1999) uses the notion of epistemic cultures to explain and contrast domain differences in knowledge-making processes. She defines epistemic cultures (1999, 1, emphasis in original) as "those amalgams of arrangements and mechanisms . . . which, in a given field, make up *how we know what we know*. Epistemic cultures are cultures that create and warrant knowledge." This dovetails neatly with Kling's (2004) accounts of publishing practices within high-energy physics. He has described how membership of, and authorship responsibilities within, large collaborations (such as DZero and BTeV at Fermilab) are tightly controlled by the collective. In similar vein, Biagioli (2003) has furnished a detailed account of the bases on which the CDF (Collider Detector at Fermilab) Collaboration's members are included in the Standard Author List. These localized norms and practices are part of what is denoted by the phrase "epistemic culture," what van House (2004, 27) terms "the complex texture of knowledge as practiced." Many physicists work in huge, distributed collaborations comprising hundreds of individuals. These largely self-regulating collectivities have local rules and detailed procedures governing participation, internal review, and publication practices. As a result, high-energy (and other) physicists routinely post their papers on e-preprint archives/institutional repositories, and coauthorship is the norm—often involving scores or hundreds of individuals.

The epistemic culture of high-energy physicists (as far as media use is concerned) is substantively different from other scientific disciplines, such as chemistry, where online posting/archiving of preprints is deemed unacceptable and adherence to the Ingelfinger rule—which prohibits the publication of previously posted material (see www.nejm .org/hfa/ingelfinger.asp for a discussion in respect of the *New England Journal of Medicine* [*NEJM*])—remains widespread. Because many of the premier chemical journals refuse to publish articles that have been posted on the web, Elsevier, the leading commercial publisher in the discipline, launched an electronic preprint server, ChemWeb, in an effort to stimulate the kinds of practices pioneered by the physicists. Judging by Warr's (2003) interim evaluation of the Elsevier service and its subsequent discontinuation in May 2004 (see http://preprint.chemweb .com/CPS/show/), there is still considerable resistance to the idea of preprint servers within the ranks of chemists (but not quantitative biologists who have just launched the q-bio archive). In part, this may be due to the dominant role played by the American Chemical Society and *Chemical Abstracts* within the chemistry community; in part, it may be a function of differences in the biorhythms of the two disciplines. Physicists came up with *Physics Letters* and then e-preprint servers as ways of bringing their results to the peer community as quickly as possible: rapid publication matters more in physics than chemistry.

At the risk of stating the obvious, the epistemic culture of the high-energy physics community is far removed from the world of humanities scholarship (e.g., literature and literary criticism) where the scholarly monograph is revered and the phenomena of fine-grained division of "cognitive labor" (Kitcher 1993, 303), "collective cognition" (Giere 2002), and hyperauthorship are virtually unknown. In a brace of longitudinal studies, we (Cronin, Shaw, and La Barre 2003, 2004) have shown how the intensity of informal collaboration (revealed, for instance, in the number of trusted assessors mentioned in acknowledgments) and degree of coauthorship differ significantly across disciplines (chemistry, psychology, and philosophy were three exemplary literatures we surveyed) and also how these measures have changed over time (see also Cronin 1995). Laband and Tollison (2000) have provided novel insights into coauthorship and collaboration in economics and biology journals for the period 1950–1994. In extreme cases such as high-energy physics and large-scale clinical trials studies, massive collaborations can result in "the erasure of the individual as an epistemic subject" (Knorr Cetina 1999, 166), which is roundabout way of saying that it is well nigh impossible for an outsider to determine exactly who did what (see also Davenport and Cronin [2001] on credit allocation in group work). These are not typically issues that occupy the minds of humanities scholars—an admittedly coarse-grained if convenient portmanteau term—for whom sole authorship is still overwhelmingly the norm. The problems of attribution and accountability that have been extensively documented in the physical and biomedical sciences do not occur in the humanities, where text and author are tightly coupled; where the process of inscription implies intimacy with one's materials. Despite the fact that "[e]xperience tells us that our creative practices are largely derivative, generally collective and increasingly corporative and collaborative" (Jaszi and Woodmansee 2003, 195), a culture of individualism remains strong in the humanities.

DISCIPLINARY DISCOURSES

Disciplines and epistemic cultures differ in a variety of ways, at once obvious and highly subtle. Media use is a case in point, academic discourse another. There are many genres and subgenres of academic writing reflecting the different ways scholars produce, evaluate, and present evidence. As has been noted, "if one wishes to produce discourse successfully within a particular field, one must observe the forms and formalities of that field" (Thompson 1991, 20). Or, as Najman and Hewitt (2003, 77) phrase it: "[D]ifferent fields or disciplines have different profiles in terms of their research publications, their pattern of citations and their overall research quality." The ways in which arguments are constructed and rhetorical tropes deployed by different academic tribes have been carefully documented (Bazerman 1988), and Swales (1998) has produced a "textography" of a university building, *Other Floors, Other Voices*, capturing the discursive lives of its different occupants. Academic writing is anything but standardized: just pick up any two issues of, say, *JAMA* (*Journal of the American Medical Association*) and the *PMLA* (*Proceedings of the Modern Language Association*) and the structural (the chunking and sequencing of the text) and stylistic differences (disputatiousness, subjectivity) will be nothing less than striking. Nor is writing something that one does as an afterthought to the serious business of thinking and experimentation; as Hyland (2000, 3) argues, "discourse is socially constitutive rather than simply socially shaped; writing is not just another aspect of what goes on in the disciplines, it is seen as *producing* them." And, of course, the forms of writing and communicative conventions favored by different epistemic cultures will have a bearing on how ICTs, specifically electronic publishing technologies, are adopted and co-opted.

Discursive and publication practices are shaped by the way the academic reward system is operationalized from one discipline to another, a subject we'll return to in chapter 6. Publishing practices differ; for example, disciplines such as molecular biology "follow a pattern characterized by a large number of relatively short papers with joint authorship, frequently appearing in highly cited journals," whereas philosophy is "characterized by the publication of monographs and longer papers with relatively few articles appearing in journals which themselves tend to be infrequently cited" (Najman and Hewitt 2003, 77). To take another example, an assistant professor of French or German will typically be expected to have published at least one sole-authored scholarly monograph with a reputable university or commercial press when coming up for tenure, while his opposite number in high-energy physics might achieve promotion/ tenure for contributions to a number of multiauthored papers. In the former case, the candidate will likely have written every word of the book himself; in the latter, the nominal "author" may have written very little, if any, of the e-preprints/journal articles that constitute his promotion dossier. This illustrates the fundamental difference between "authorship" and "contributorship," a distinction that reflects not just variations in the stylistics of academic writing but deep differences in the culturally validated, knowledge-creation processes sanctioned by different epistemic cultures. Our physicist may actually have "written" little of the papers to which he has contributed, but the collaborative, transparent, and inherently trusting nature of the high-energy physics research world ensures that he will have made a material contribution to the collective effort: "a member who has paid his or her dues through labor becomes an author" (Biagioli 2003, 270).

There is, of course, no in-principle reason why our stereotypical humanist could not break out of the monograph mold and submit a series of thematically linked journal articles, online postings, and/or presentations as his dossier. As John Unsworth (2003), a professor of English, library school dean, and founding editor of the electronic-only journal *Postmodern Culture*, has remarked, the solution to the "crisis in scholarly publishing in the humanities"—a crisis foregrounded by Stephen Greenblatt in his year as MLA president (see http://chronicle .com/jobs/2002/07/2002070202c.htm for his open letter to the MLA membership)—is "to accept several scholarly articles in place of a book." Unsworth practices, or exemplifies, what he preaches: "I . . . got tenure in a top-ranked English department without a book—tenure was based, instead, on article-length pieces, many of which were published electronically, and on applied research (in electronic publishing)." He is not a lone pioneer; Wittenberg (1998) has described a discipline-based website created by Columbia University Press, a "venue in which international-affairs scholars can disseminate their best work at different stages and in different forms . . . making available creative, cutting-edge scholarship quickly and widely to a large community of users." Many other possibilities exist.

But things are not altogether straightforward. Within parts of academia there is considerable inertia, and old practices die hard. There is no insurmountable technological reason why humanists should not produce article-length output for publication in electronic journals or deposit in institutional archives, or experiment with novel technologies. For instance, in assessing historians' use of e-books, Musto has been quoted as follows (see *Library Journal Academic Newswire*™: The Publishing Report for July 31, 2003): "In e-publishing the classic image of the monograph falls apart." The article continues, "The idea of the monograph, a single copy of an expensive, static book, is slowly being replaced . . . by the notion of a "database" of cross-searchable, highly enhanced works that offer clear advantages, such as links to source documents and multiple, remote user access. . . . [R]esearchers and students, who generally use portions of monographs for their work, are finding that e-books better support much of their current behavior and offer much promise for expanding the way scholarship is communicated." A good illustration of change from within has been the establishment of NINES (Networked Infrastructure for Nineteenth-Century Electronic Scholarship; see http://faustroll.clas.virginia.edu/nines/user/view/2), both a group and a project, spearheaded by Jerome McGann, to provide the means for peer-reviewed online publishing of Romantic and Victorian British and American scholarship in nineteenth-century studies.

Humanists are no less competent than physicists, mathematicians, and other scholars when it comes to posting their papers on preprint servers. Rather, the publication practices of most humanists are shaped by a set of institutionally embedded norms and material practices that are not found in other epistemic cultures. For example, originality is defined differently in different settings: "humanists valued the use of an original approach and new data most frequently; historians privileged original approaches above all other forms of originality; while social scientists emphasized the use of a new method" (Guetzkow, Lamont, and Mallard 2004, 201). For many humanists, at least those who populate promotion and tenure committees, the monograph remains the most revered mode of publication, and the integrity of a scholar's work is guaranteed by the reputation of the sponsoring press and the associated peer review procedures. This publication model is inscribed in many of the promotion and tenure guidelines used in North American universities, and is very far removed from the "post-traditional communitarian structures" (Knorr Cetina 1999, 165) embraced so enthusiastically by high-energy physicists. The dogged focus on publication form (scholarly monograph, etc.) that defines the status quo is unlikely to persist in perpetuity. As Clark (2003, 156) has observed, "[B]undling information into preset, pretagged physical packages . . . may thus become less and less crucial, as users learn to soft assemble resources pretty much at will, tailored to their own specific needs. One result of this may be the gradual erosion of the firewalls that currently separate various documents and sources of information."

MATERIAL PRACTICES

But there is another wrinkle. For many scholars in the humanities, securing a contract with a university press may be critical to launching their careers. Yet, as Regier (2003) has shown, the university press, it-self, is something of an endangered species. From 1980–2000, the out-put of university presses grew faster than the market, and so they began to target the trade book sector as a source of supplementary revenue. During the same period, faculty in U.S. universities grew only 65 per-cent. In 2000, university presses produced 31 million books, of which libraries bought 5 million. In recent years, sales of the average scholarly monograph produced by university presses have fallen, roughly speak-ing, from 1,500 to 200 units. Which brings us back to the ARL statis-tics. Library budgets for monographs are declining and fewer scholarly monographs are being purchased. And this is happening at a time when, according to a Modern Language Association report, *The Future of Scholarly Publishing* (Ryan et al. 2002), competition for faculty posi-tions is intensifying and the emphasis on monographic publishing is in-creasing. Proposals for reform of the status quo have been put forward by Davidson (2003).

Additionally, there are fewer publishing outlets for scholarly (as op-posed to trade) books than previously. Junior faculty in some disciplines now find themselves between a rock and a hard place. Given the pres-ent situation, there is, arguably, a need for (a) alternative forms of schol-arly expression and (b) greater emphasis on the quality rather than quantity of an individual's work. In analyzing the predicament facing humanities scholars, the MLA report (Ryan et al. 2002, 180) wonders if "our local and professional practices [are] encouraging the best work of individuals?" Note that the report did not comment on the capabilities of electronic publishing technologies, but on "local and professional practices," precisely the point made repeatedly by Kling and his col-leagues.

The diversity of epistemic cultures within the academy translates into a wide range of distinctive practices. Chemists, as already noted, are disinclined to submit their papers to preprint services, while many physicists and astronomers consider it perfectly normal practice to post their findings on the web. In fact, the use of e-print archives varies strik-ingly by discipline, according to a study by Lawal (2002). She sent a web questionnaire to a random sample of 473 scholars and researchers in North American colleges and universities. I should note in passing that the sample does not appear to have been weighted by size of disci-pline or professional status. Further, we are not told what the response

rate was, nor are we provided with the raw data from which the per-
centages are derived. Of those responding, 18 percent used e-prints; 82
percent did not. According to Lawal, the use of e-print archives ranged
as follows: physics/astronomy, 52 percent; mathematics/computer science, 29 percent; engineering, 7 percent; cognitive science/ psychology,
7 percent; biological science, 4 percent; chemistry, 0 percent. Even allowing for concerns about reliability, these figures reinforce the anec-
dotal evidence that adoption rates for e-print archiving vary appreciably
from discipline to discipline.

Different modalities of publication hold primacy in different disciplines. Refereed conference presentations in computer science are as, if
not more, valued than peer-reviewed journal articles. Economics has
been described as "probably the most rigid discipline in terms of the hi-
erarchy of journals that count" (La Manna 2002), a discipline, moreover, for which self-archiving "as a strategy to free access to peer-
reviewed research . . . is, and can easily be proved to be, a total non-
starter." In linguistics, journal article publication is the norm. Historians, to take another example, rely heavily on monographs (Collier
1999); writing and referencing scholarly tomes are central to historiography. However, there are signs that this may be beginning to change.
Crane (2003, A37) believes that the growth in small-scale digitization
projects means that "[h]istorians won't be building their work around
the assumption that paper based projects are the be-all and end-all."
And historians typically fly solo: the kinds of coauthorship practices
that dominate scientific publication are unheard of, if not unimaginable,
in history. The lone scholar may be an anachronism in most scientific
and many social scientific disciplines, but that is not the case in the humanities.

Collaboration—for which coauthorship is the most visible and hand-
iest indicator—is established practice in both the life and physical sciences, reflecting the industrial scale, capital-intensiveness and complexity of much contemporary scientific research. But the "standard
model of scholarly publishing," one that "assumes *a* work written by *an*
author" (Cronin 2002, 559, italics added), continues to hold sway in the
humanities. As already noted, the behaviors of high-energy physicists,
a specialty group studied closely by, amongst others, Kling (e.g., Kling,
McKim, and King 2003), Knorr Cetina (1999), and Traweek (1992), are
very different from those of the average professor of history or French.

In the humanities, career advancement depends on conformance to an essentially individualistic model of scholarly production. Relatively few primary electronic sources will be referenced because of lack of perceived respect for online sources and also because much potentially relevant material is simply not available in electronic form; material such as medieval manuscripts housed in a monastery or as yet uncatalogued Third World government archives—see Graham (2002) on historians' use of electronic resources. This, of course, is a simplification. There are many electronic publishing pioneers and enthusiasts to be found in the humanities—Gregory Crane, Jean-Claude Guédon, Jerome McGann, and John Unsworth being four notable examples. But broadbrush differences in scholars' material practices do, nonetheless, exist.

TRUSTING RELATIONS

By way of contrast, physicists do not mandate sole-authored output. Nor does the research monograph have a place in their world. Collaboration and coauthorship are necessarily the norm. Posting, reading, and referencing e-preprints is established practice. Theirs, too, is a much more trusting system, and it works partly because of what Traweek (1992, 117) describes as physicists' "agnostic evaluation of other physicists"—that is to say, the high levels of self-scrutiny and internal vetting that characterize this particular epistemic culture. The social bases of trust in science have been explored by two of the authors cited in the opening paragraph of this chapter, Robert Merton and Steven Shapin. Merton (1976) was instrumental in positing a set of governing norms for scientific conduct, and Shapin (1994) has detailed how in the seventeenth century the trustworthiness of scientific claims was explicitly linked to the personal trustworthiness (embodied in one's social standing or gentlemanliness) of those making the judgments. The conventions for evaluating research may have changed in the last few centuries, but trust, a manifestation of the "normative ghost in the scientific machine" (Cronin 2004), and peer review, the instrument for ensuring trustworthiness, remain central to the conduct of science in general (Davenport and Cronin 2000) and, specifically, to the smooth functioning of the primary communication process: "If it [an article] is cited many times as the basis for other research, then it gains trustworthiness. The laboratories and scientists who have been cited successfully in turn become validating voices for other articles when they cite them" (Schlossberg 1999, 69). The ways in which trust is constituted and sustained vary across epistemic cultures, and these variations will influence developments in digital media and publication technologies.

In some domains, especially biomedicine, the issues of fraud and honorific authorship have been widely discussed in recent years. Horton (1998, 688) in a *Lancet* editorial, has spoken of "the shattered system of academic reward and its symptom, broken rules of authorship." Such has been the level of concern about excessive (and unwarranted) coauthorship practices that there has been a move to retire the term "author" and replace it with "contributor." By listing all the contributors to a study/paper and specifying their particular inputs (e.g., experimental design, statistical analysis), it is argued, confusion about who did what will be removed and credit can be allocated equitably (Rennie, Yank, and Emanuel 1997). The single author may be dead in biomedicine, but it is unlikely that the radical contributorship model will find easy favor in other epistemic cultures, particularly those where the act and craft of writing are so intimately associated with the scholarly end product, where "thinking *via* the act of writing" (Clark 2003, 5) is a fact of life.

As might be expected from the argument thus far, the way in which trust is instantiated in institutional arrangements for peer review varies across epistemic cultures. The physicists who read and cite papers posted on Paul Ginsparg's emblematic e-print archive, arXive.org, are not rejecting conventional peer review (e-preprints, in any event, migrate in time to refereed journals), but their confidence in the reliability of the collective arrangements for validating new knowledge claims in their domain is such that they can reliably draw upon the work of their peers *before* it comes to rest in the relevant journal of record. The mechanisms that facilitate social trust building in their professional culture entail relatively little risk; parenthetically, they instantiate Merton's classical norms. Things are otherwise in the world of chemistry, where researchers' material practices and also modes (and scale) of collaboration are quite different from those in high-energy physics. In addition, the voice and views of the leading professional body, the American Chemical Society (also the leading publisher), play a role in shaping researchers' writing and publication behaviors (see, for instance, the ACS style guide at www.oup-usa.org/sc/0841234620/0841234620_1.html).

As Hilf (2003), a physicist, noted in relation to the growth in the number of papers being posted on arXive.org, there will never be "saturation such that all papers will go this way, since in different fields *culture* and *habits* and requirements are different" (italics added).

When it comes to the humanities, the cultural differences are even more pronounced. Here, trust is less personally rooted than institutional in character. The community of scholars is more dispersed and, in some respects, less self-knowing. Humanities scholars typically work alone rather than as members of a large collaboration. It is the reputation of academic publishers and journals, and the presumed rigor of the associated peer review procedures, that provide the foundations of the system's trustworthiness rather than the intensity of personal/sociocognitive ties that exist between individuals or distributed workgroups/research teams.

The limitations of the peer review system (at least in the field of cultural studies) were famously exposed by the so-called Sokal hoax (Sokal and Bricmont 1999). The publication of Sokal's parodic article ("Transgressing the Boundaries: Toward a Transformative Hermeneutics of Quantum Gravity") in the journal *Social Text* shows just how important mastery of discursive moves and tactics is in certain academic contexts. That a member of another tribe (a physicist) could so deftly ape the convoluted writing style favored by some postmodernists, and also persuade the journal's referees of the merits of his intentional nonsense, is most revealing. In their critical review of *soi-disant* postmodern scholarship, Sokal and Bricmont (1999, 6) make an important observation: "There are many different degrees of abuse. At one end, one finds extrapolations of scientific concepts, beyond their domain of validity, that are erroneous but for subtle reasons. At the other end, one finds numerous texts that are full of scientific words but entirely devoid of meaning." Paradoxical though it sounds, literal meaninglessness seems to be tolerated in some epistemic cultures or subcultures—even, lamentably, classics (see Hanson, Heath, and Thornton [2001] for an insight into the prevalence of pseudo-scholarship).

PLURALITY AND PLASTICITY

Resistance to "scholarly skywriting" (Harnad 1990) stems from concerns about bypassing the long-established, albeit imperfect, peer review process—the lynchpin of the quality assurance system. New modes of posting and self-publishing are seen as potentially corrosive to the bases of trust and integrity traditionally associated with scholarly publishing. It is, of course, wholly mistaken to equate open-access initiatives (e.g., Public Library of Science, Open Society Institute) with subversion of established peer review practices. However, the new forms of web-based and electronic publishing do make it possible to bring one's work directly and swiftly to a much wider audience by circumventing the normal editorial filters and controls. Liberationists argue that open (or deep) peer review has much to commend it, quite apart from faster production and distribution, as shoddy work and dubious findings will be subjected to the full glare of public scrutiny and, thus, will revealed for what they are. Harnad (1979) has long been vocal in the vanguard of the drive for open peer commentary and the promotion of "creative disagreement" in scientific communication and debate. Proponents of open access argue that the web invites public inspection and fosters transparency. Odlyzko (2002, 9) believes that pointers to web-based resources (implicit recommendations, in effect) constitute "a form of peer review." Establishmentarians, on the other hand, feel that the untrammeled flood of papers and postings on the web will make it well nigh impossible to sort the scientific wheat from the proliferating chaff. Battle has been joined.

But it is not a strict dichotomy. It may be more helpful to think of peer review as a spectrum, moving from heavy (anonymous, double-blind) peer review through peer review "lite" to open peer review (I label the latter "fear review" because one's work is exposed to 360-degree scrutiny, which may well be more nerve-racking than having it read unnoticeably by only two or three peers). I use the term "dear review" to signify that referees and reviewers are sometimes paid for their services and provided with pecuniary incentives to turn around papers expeditiously. In addition, Kling (2004) has coined the term "career review" to describe the kind of internal screening that takes place in the context of guild publishing enterprises, such as an academic department's occasional, working, or technical paper series. Materials added to such a series (whether print based or electronic) are not peer reviewed as such; admission instead depends on one's institutional affiliation and acceptance by the local authorial/scholarly community. In exceptional cases, peer review is anything but peer based. In law, one might say, it is a case of the lunatics running the asylum.

Law reviews are, in fact, edited by law school students not by the field's leading scholars. (The rankings of law reviews/law journals are largely a reflection of law school rankings.) Here the apprentices critique the masters' submissions. This idiosyncratic model (demonstrating, I suppose, that the child is the father of the man) seems to work despite its inversion of established practice.

The world of research and scholarship comprises many disciplines and a mélange of epistemic cultures. This heterogeneity of behaviors, values, and practices means that ICTs are deployed differently, hence the title of Kling and McKim's 2002 paper, "More Than a Matter of Time." Of course, it is not just a matter of time before all disciplines reach the "tipping point" (Gladwell 2002), before they adopt the arXive.org (or some other more or less standardized) approach to electronic publishing/archiving. The picture is much more complicated, and the reasons why, and ways in which, researchers commit to particular e-publishing regimes will be a function of a field's sociocognitive structure, history, normative character, and institutional circuitry—not just its metabolic rate.

Even if we accept that there are significant differences in the ways in which scholars create and disseminate new knowledge, and even if we allow that scholars' material practices are shaped by the prevailing epistemic culture, is it not the case, you might ask, that the resistors—the e-publishing laggards—will eventually come on board as web-based publishing technologies become ubiquitous and even easier to use? Will conservative humanists be able to resist the lure of electronic publishing and digital archiving, given the nature of the job market, the demands of the academic reward system, and the grim economic realities of monograph publishing? Will the principled resistance to piecemeal publishing crumble in a generation or so, as the ancien régime fades and a new breed of hybrid scholar moves through the ranks, dominating promotion and tenure committees and progressively rewriting/reinterpreting the guidelines? If high-energy physicists are the early adopters of web-based self-archiving, why not simply see the majority of historians, philosophers, and literature professors as the late adopter population? Perhaps it is, in fact, just a matter of time. But this is to confuse infrastructural with cultural change. Much (not all) scholarly publication will migrate to the web—the default platform of choice—but the ways in which ICTs are used to communicate with one's peers and disseminate one's ideas will continue to mirror underlying differences in epistemic cultures and disciplinary value systems.

DISRUPTIVE TECHNOLOGIES

This is a good point to step back from the minutiae of tribal behaviors and look at the bigger picture—at the structural dynamics of the scholarly communication ecosystem. Christensen (2000) has described with well-documented examples from a range of industrial and service sectors how disruptive technologies can lead to the undermining and eventual demise of a product or service that dominates the market. He chronicles how many companies have engaged in "sustaining innovation," adding, in other words, more and more layers of sophistication or feature-richness to a product for an important and established customer segment; he terms this "performance oversupply." One consequence is that those who lack the financial means or skills to benefit from a particular product or service are effectively locked out of the market. According to Christensen, in Type I disruptions, a competitor launches a product that is comparatively simple and/or cheaper to use and thus of potentially broad appeal. The innovator in this case is providing the market with a new product; examples he cites include PCs, the telephone, and personal financial management software. Each of these technological innovations created a new market for those who could not access mainframes, for those who could not use/get to the telegraph office, and those who couldn't afford tax consultants. Type II disruptions, on the other hand, occur when a supplier or manufacturer targets the soft underbelly, or low end, of the existing market with an appealing alternative to the prevailing model. Examples of this are discount retailing, steel mini-mills, and no-frills airlines. Over time, the once-dominant players begin to see their market share diminish as their customers migrate to the products of interlopers.

Odlyzko (2002, 10) has applied Christensen's displacement model to scholarly communication, arguing that electronic publishing exhibits three important characteristics of disruptive technologies (underperforming existing products; enabling new applications; rapid performance improvement). He contrasts macrolevel trends, and concludes that the writing is all but on the wall for the Old World order. The first is annual attrition in print journal subscriptions rates (in the 3–5 percent range); the second is a continuous growth in online access (he provides multiyear visitor/host data on a variety of websites to support his contention, along with download statistics on arXive.org articles). The Old World order is waning and the New World order is waxing. We may be some way off the proverbial tipping point, but, as he goes on to say (2002, 8), "the attraction of a much greater audience on the web, and the danger that anything not on the web will be neglected, are likely to become major spurs to scholars' migration of their work online." Supporting evidence for his thesis comes from the two sets of ARL statistics mentioned earlier.

The appeal of the web, for both producers and consumers of texts, is undeniable, if sometimes exaggerated: access to wider readership, potentially higher citation rates, shorter time to market, bypassing of filters and barriers, dynamic updating of content, local authorial control, self-branding opportunities, informal/open peer review, ease of use, linking to multimedia resources, and control over presentation. As more and more publication and dissemination activity moves to the web, the resistors will be forced to reconsider some of their engrained practices. For example, it is hard to imagine that the world of monograph publishing in the humanities will not experience significant change in the next decade, that early experimentation will not stimulate further exploration of new modalities of scholarly communication by the various stakeholders—individual scholars, universities, consortia, university presses, institutional and commercial publishers. The attractions of online publishing and posting, combined with the inefficiencies of the traditional scholarly publishing model, make transdisciplinary innovation irresistible, as the MLA and other organizations are increasingly aware. In a *PMLA* editorial, Alosnso (2003, 222) suggests that "the association could become the electronic repository of manuscripts recommended by divisional executive committees, thereby contributing to the dissemination of research judged significant by some of the best scholars in the field."

Advances in electronic publishing will certainly not erase the cultural differences between disciplines. Instead, deep-rooted normative and behavioral differences will stimulate the creation and adoption of yet other new ways of doing business. Odlyzko is right to focus on the high-level trends. The large-scale shift from print to web-based publishing, on the one hand, and the gradual rise of the open-access movement on the other, are two trends that will change fundamentally the balance of power in the scholarly communication marketplace, but it is still not clear what form self-archiving will take. As Harnad (2003) put it, "The reason institutional self-archiving is more likely to speed up self-archiving and to generalize it across disciplines is that researchers and their institutions both share the benefits of the impact of their research output, whereas researchers and their disciplines do not. It is not the discipline that exercises the incentive of the 'publish-or-perish' carrot-and-stick on researchers, it is their research institutions." It is unwise, however, to assume that a single model will emerge and dominate the landscape. Disciplines will continue to view and co-opt ICTs in different ways, and the upshot will be a kaleidoscope of initiatives and local adaptations. Indeed, the long-term effect of widespread disciplinary differences coupled with a rich array of technological possibilities will be the formation of a much more heterogeneous and dynamic publishing ecosystem than before, one that supports a multiplicity of epistemic cultures; "an opportunity environment" in which various species "co-evolve" (Moore 1996, 11, 16), resulting in new hybrids with new tolerances.

CONCLUSIONS

There is a voluminous and many-sided literature—speculative, advocacy, and research based—on developments relating to scholarly communication and academic publishing. Statistical data and insightful analysis on the economics of scholarly and commercial publishing are abundant (Noll and Steinmuller 1992; McCabe 2002), and the professional library literature is replete with descriptions and evaluations of experimental online services and products. In addition, there is a growing body of evidence to suggest that being online increases the visibility of one's work (e.g., Lawrence 2001). But back to Merton once again, and this time the norm of communism, the idea that scientists are impelled to share and make freely available the fruits of their intellectual labor. What better way to achieve optimal exposure than by posting one's work on the web or submitting one's papers to open-access journals, and in so doing making access to new knowledge as simple as possible for the wider, international community? The corollary of this is that greater exposure ultimately leads to greater recognition (all things being equal).

As online and web-based publishing become increasingly common, altruism and self-interest dovetail: "Whether we like it or not, scholars are engaged in a "war for the eyeballs'" (Odlyzko 2002, 18)—a kind of "vanity fair" (Franck 1999), if you will. One result is that new measures of "scholarly salience" and "presence density" are emerging (Cronin 1999b, 953), complementing traditional bibliometric impact measures. Branding, competition, and vanity will help propel the open-access publication movement, championed by organizations such as PLoS and legitimated by the introduction in June 2003 of the Public Access to Science Act, a bill that would "exclude from copyright protection works resulting from scientific research substantially funded by the Federal Government" (see www.theorator.com/bills108/hr2613.html). In concrete terms, we can expect to see more initiatives such as MIT's DSpace—a digital repository to capture, distribute, and preserve the institution's intellectual output and make it accessible via the web to the general public (see www.dspace.org/)—and University of California Press eScholarship Editions, which makes hundreds of its titles available to the public at no cost (see http://escholarship.cdlib.org/ucpress/). But the particulars of specific initiatives and proposals matter less than changes in the underlying dynamics of scholarly publication.

The picture that emerges is of a very complicated ecosystem, the contours of which are still blurred. Perhaps it would be more helpful to think in terms of an *ethological* approach focusing on the behavioral characteristics of the primary stakeholder groups. What is missing from the larger literature is a compelling analysis of the structural dynamics of the scholarly communication marketplace, one that focuses upon the array of stakeholder relations, technological drivers, and competitive forces (and their interactions) that are reconfiguring the ecosystem. One looks in vain for the kind of high-level, strategic analysis a Michael Porter might provide (Porter 1998, 2001), particularly one informed by the results of sociotechnical systems studies (Bohlin 2004) and meticulous ethnomethodological investigation.

For most of the twentieth century, there was one medium of scholarly publishing: print. At the same time, there were a few well-established genres of academic writing, the monograph and journal article being the dominant textual forms across disciplines. In such an environment, it made sense to talk about "*the* primary communication system" or "*the* scholarly publication system." Monothetic thinking was not out of place. Today there is a plurality of media and genres; scholars can publish, distribute, post, and archive their research in a variety of ways. New publishing modalities are emerging, new forms of collaboration are establishing themselves, and new approaches to peer review are being trialed. Disciplines have different biorhythms; the pace of new knowledge creation is faster in some fields than others, the need for interaction and feedback more pressing. Different approaches and solutions will be adopted at the local level. The present environment allows communication channels and information resources to be matched more effectively with the cultural characteristics and needs of epistemic communities. In short, the big picture has to be considered along with a collection of miniatures, if the dynamics of scholarly communication and publication are to be properly grasped.

REFERENCES

Alonso, C. J. (2003). Editor's column: Having a spine—Facing the crisis in scholarly publishing. *Proceedings of the Modern Language Association*, 118(2), 217–223.

Ayris, P. (n.d.). *Scholarly Communications Crisis: Setting the Context*. Available at www.ucl.ac.uk/scholarly-communication/ADVOC.PPT.

Bazerman, C. (1988). *Shaping Written Knowledge: The Genre and Activity of the Experimental Article in Physics*. Madison: University of Wisconsin Press.

Becher, T. (1989). *Academic Tribes and Territories: Intellectual Enquiry and the Culture of Disciplines*. Milton Keynes: Open University Press.

Biagioli, M. (2003). Rights or rewards? Changing frameworks of scientific authorship. In: Biagioli, M. and Galison, P. (Eds.). *Scientific Authorship: Credit and Intellectual Property in Science*. New York: Routledge, 253–279.

Bishop, A. and Star, S. L. (1996). Social informatics for digital libraries. In: M. E. Williams (Ed.). *Annual Review of Information Science and Technology*, 31. Medford, NJ: Information Today, 301–402.

Bohlin, I. (2004). Communication regimes in competition: The current transition in scholarly communication seen through the lens of the sociology of technology. *Social Studies of Science*, 34, 365–391.

Christensen, C. M. (2000). *The Innovator's Dilemma: When New Technologies Cause Great Firms to Fail*. Cambridge, MA: Harvard Business School Press.

Clark, A. (2003). *Natural-born Cyborgs: Minds, Technologies, and the Future of Human Intelligence*. Oxford: Oxford University Press.

Collier, B. (1999, February 5). Preserving the central role of the monograph. *Chronicle of Higher Education*, A56.

Crane, D. (1972). *Invisible Colleges: Diffusion of Knowledge in Scientific Communities*. Chicago: University of Chicago Press.

Crane, G. (2003, September 5). Quoted in: Brock, R. How digital hobbyists are changing scholarship. *Chronicle of Higher Education*, A37–38.

Cronin, B. (1982). Invisible colleges and information transfer: A review and commentary with particular reference to the social sciences. *Journal of Documentation*, 38, 212–236.

Cronin, B. (1989). Research libraries: An agenda for change. *British Journal of Academic Librarianship*, 4(1), 19–26.

Cronin, B. (1995). *The Scholar's Courtesy: The Role of Acknowledgement in the Primary Communication Process*. London: Taylor Graham.

Cronin, B. (1999a, October 15). Will e-publishing save academic research? *Chronicle of Higher Education*, A25.

Cronin, B. (1999b). The Warholian moment and other proto-indicators of scholarly salience. *Journal of the American Society for Information Science*, 50(10), 953–955.

Cronin, B. (2001a). Bibliometrics and beyond: Some thoughts on web-based citation analysis. *Journal of Information Science*, 27(1), 1–7.

Cronin, B. (2001b). Knowledge management, organizational culture and Anglo-American higher education. *Journal of Information Science*, 27(3), 129–137.

Cronin, B. (2002). Hyperauthorship: A postmodern perversion or evidence of a structural shift in scholarly communication practices? *Journal of the American Society for Information Science and Technology*, 52(7), 558–569.

Cronin, B. (2004). Normative shaping of scientific practice: The magic of Merton. *Scientometrics*, 60(1), 41–46.

Cronin, B., Shaw, D., and La Barre, K. (2003). A cast of thousands: Coauthorship and subauthorship collaboration in the 20th century as manifested in the scholarly journal literature of psychology and philosophy. *Journal of the American Society for Information Science and Technology*, 54(9), 855–871.

Cronin, B., Shaw, D., and La Barre, K. (2004). Visible, less visible, and invisible work: Patterns of collaboration in twentieth century chemistry. *Journal of the American Society for Information Science and Technology*, 55(2), 160–168.

Cummings, A. M., Witte, M. L., Bowen, W. G., Lazarus, L. O., and Ekman, R. H. (1992). *University Libraries and Scholarly Communication. A Study Prepared for the Andrew W. Mellon Foundation*. Washington, DC: Association of Research Libraries.

Davenport, E. and Cronin, B. (2000). The citation network as a prototype for representing trust in virtual environments. In: Cronin, B. and Atkins, H. B. (Eds.). *The Web of Knowledge: A Festschrift in Honor of Eugene Garfield.* Medford, NJ: Information Today, 517–534.

Davenport, E. and Cronin, B. (2001). Who dunnit? Metatags and hyperauthorship *Journal of the American Society for Information Science and Technology*, 52(9), 770–773.

Davidson, C. N. (2003, October 3). Understanding the economic burden of scholarly publishing. *Chronicle of Higher Education*, B7–10.

Franck, G. (1999). Scientific communication—A vanity fair? *Science*, 286, 53, 55.

Friedlander, A. and Bessette, R. (2003). *The Implications of Information Technology for Scientific Journal Publishing.* Arlington, VA: National Science Foundation.

Garvey, W. D. and Griffith, B. C. (1971). Scientific communication: Its role in the conduct of research and creation of knowledge. *American Psychologist*, 20(1), 349–362.

Giere, R. N. (2002). Distributed cognition in epistemic cultures. *Philosophy of Science*, 69(4), 637–644.

Gladwell, M. (2002). *The Tipping Point: How Little Things Can Make a Big Difference.* New York: Little, Brown.

Graham, S. R. (2002). Historians and electronic resources: Patterns and use. *Journal of the Association for History and Computing*, 5(2). Available at http://mcel.pacificu.edu/JAHC/JAHCV2/ARTICLES/graham/graham.html.

Guédon, J.-C. (2003). Open access archives: from scientific plutocracy to the republic of science. *IFLA Journal*, 29(2), 129–140.

Guetzkow, J., Lamont, M., and Mallard, G. (2004). What is originality in the humanities and the social sciences? *American Sociological Review*, 69, 190–212.

Hanson, V. D., Heath, J., and Thornton, B. S. (2001). *Bonfire of the Humanities: Rescuing the Classics in an Impoverished Age.* Wilmington, DE: ISI Books.

Harnad, S. (1979). Creative disagreement. *The Sciences*, 19, 18–20.

Harnad, S. (1990). Scholarly skywriting and the prepublication continuum of scientific inquiry. *Psychological Science*, 1, 342–343.

Harnad, S. (2003). Central vs. distributed archives. Posted to SEPTEMBER98-FORUM, September 10.

Hilf, E. (2003). Central vs. Distributed Archives. Posted to SEPTEMBER98-FORUM, September 8.

Horton, R. (1998). The unmasked carnival of science. *The Lancet*, 351(9104), 688–689.

Hyland, K. (2000). *Disciplinary Discourses: Social Interactions in Academic Writing.* New York: Longman.

Jaszi, P. and Woodmansee, M. (2003). Beyond authorship: Refiguring rights in traditional culture and bioknowledge. In: Biagioli, M. and Galison, P. (Eds.). *Scientific Authorship: Credit and Intellectual Property in Science*. New York: Routledge, 195–223.

Kitcher, P. (1993). *The Advancement of Science: Science Without Legend, Objectivity Without Illusions*. New York: Oxford University Press.

Kling, R. (2004). The Internet and unrefereed scholarly publishing. In: Cronin, B. (Ed.). *Annual Review of Information Science and Technology*, 38. Medford, NJ: Information Today, 591–631.

Kling, R. and Callahan, E. (2002). Electronic journals, the Internet, and scholarly communication. In: Cronin, B. (Ed.). *Annual Review of Information Science and Technology*, 37. Medford, NJ: Information Today, 127–177.

Kling, R. and McKim, G. (2002). Not just a matter of time: Field differences and the shaping of electronic media in supporting scientific communication. *Journal of the American Society for Information Science*, 51(14), 1306–1320.

Kling, R., McKim, G., and King, A. (2003). A bit more to IT: Scientific multiple media communication forums as socio-technical interaction networks. *Journal of the American Society for Information Science and Technology*, 54(1), 47–67.

Kling, R. and Swygart-Hobaugh, A. J. (2002). The Internet and the velocity of scholarly publishing. Available at www.slis.indiana.edu/CSI/WP/WP02-12B.html.

Knorr Cetina, K. (1999). *Epistemic Cultures: How the Sciences Make Knowledge*. Cambridge, MA: Harvard University Press.

La Manna, M. (2002). Presentation at an international conference, *The New Information Order and the Future of the Archive*. University of Edinburgh, March 20–23, 2002. Available at www.st-andrews.ac.uk/%7Emlm/HomePage/ConfELSSS.htm.

Laband, D. N. and Tollison, R. D. (2000). Intellectual collaboration. *Journal of Political Economy*, 108(3), 632–662.

Latour, B. and Woolgar, S. (1979). *Laboratory Life: The Social Construction of Scientific Facts*. Beverly Hills, CA: Sage.

Lawal, I. (2002). Scholarly communication: The use and non-use of e-print archives for the dissemination of scientific information. *Issues in Science and Technology Librarianship*. Available at www.istl.org/02-fall/article3.html.

Lawrence, S. (2001). Online or invisible? *Nature*, 411(6837), 521.

McCabe, M. J. (2002). Journal pricing and mergers: A portfolio approach. *American Economic Review,* 92(1), 259–269.

Meadows, A. J. (1974). *Communication in Science*. London: Butterworths.

Meadows, A. J. (1998). *Communicating Research*. San Diego, CA: Academic Press.

Merton, R. K. (1976). *The Sociology of Science: Theoretical and Empirical Investigations*. Chicago: Chicago University Press.

Moore, J. F. (1996). *The Death of Competition: Leadership and Strategy in the Age of Business Ecosystems*. New York: Harper Collins.

Najman, J. M. and Hewitt, N. (2003). The validity of publication and citation counts for sociology and other disciplines. *Journal of Sociology*, 39(1), 62–80.

Noll, R., and Steinmuller, W. E. (1992, Spring and Summer). An economic analysis of scientific journal prices: Preliminary results. *Serials Review*, 32–37.

North, D. C. (1990). *Institutions, Institutional Change and Economic Performance*. Cambridge: Cambridge University Press.

Odlyzko, A. (2002). The rapid evolution of scholarly communication. *Learned Publishing*, 15(1), 7–19.

Porter, M. E. (1998). *Competitive Strategy: Techniques for Analyzing Industries and Competitors*. New York: Free Press.

Porter, M. E. (2001, March). Strategy and the Internet. *Harvard Business Review*, 63–78.

Price, D. J. de. S. (1961). *Science Since Babylon*. New Haven, CT: Yale University Press.

Price, D. J. de. S. (1965). *Little Science, Big Science*. New York: Columbia University Press.

Regier, W. G. (2003, June 13). 5 problems and 9 solutions for university presses. *Chronicle of Higher Education*. Available at http://chronicle.com/free/v49/i40/40b00701.htm.

Rennie, D., Yank, V, and Emanuel, L. (1997). When authorship fails: A proposal to make contributors accountable. *Journal of the American Medical Association*, 287, 579–585.

Ryan, J. et al. (2002). The future of scholarly publishing. *Profession 2002*. New York: MLA, 172–186.

Schlossberg, E. (1999). A question of trust. *Brill's Content*, 2(2), 68–70.

Shapin, S. (1994). *A Social History of Truth: Civility and Science in Seventeenth-Century England*. Chicago: University of Chicago Press.

Sokal, A. and Bricmont, J. (1999). *Fashionable Nonsense: Postmodern Intellectuals' Abuse of Science*. New York: Picador.

Swales, J. (1998). *Other Floors, Other Voices: A Textography of a Small Building*. Mahway, NJ: Lawrence Erlbaum.

Thelwall, M., Vaughan, L., and Björneborn, L. (2005). Webometrics. In: Cronin, B. (Ed.). *Annual Review of Information Science and Technology*, 39. Medford, NJ: Information Today, 81–135.

Thompson, J. B. (1991). Editor's introduction. In: Bourdieu, P. *Language and Symbolic Power*. Cambridge, MA: Harvard University Press.

Traweek, S. (1992). *Beamtimes and Lifetimes: The World of High Energy Physics*. Cambridge, MA: Harvard University Press.

Unsworth, J. (2003, online). *The Crisis in Scholarly Publishing. Remarks at the 2003 Annual Meeting of the American Council of Learned Societies*. Available at www.iath.virginia.edu/~jmu2m/mirrored/acls.5-2003.html.

van House, N. (2004). Science and technology studies and information studies. In: Cronin, B. (Ed.). *Annual Review of Information Science and Technology*, 38. Medford, NJ: Information Today, 3–86.

Warr, W. A. (2003). Evaluation of an experimental chemistry preprint server. *Journal of Chemical Information and Computer Science*, 43(2), 362–373.

Wellcome Trust (2003). *Economic Analysis of Scientific Research Publishing: A Report Commissioned by the Wellcome Trust*. Available at www.wellcome.ac.uk/en/1/awtpubrepeas.html.

White, H. S. (1977). The economic interaction of scholarly journal publishing and libraries during the present period of cost increases and budget reductions: implications for serials librarians. *Serials Librarian*, 1(3), 221–230.

Wittenberg, K. (1998). A new model for scholarly publishing. *Journal of Electronic Publishing*, 3(4). Available at www.press.umich.edu/jep/03-04/ciao.

Chapter Three

Hyperauthorship

Classic assumptions about the nature and ethical entailments of authorship (what I call the standard model) are being challenged by developments in scientific collaboration and multiple authorship. In the biomedical research community, multiple authorship has increased to such an extent that the trustworthiness of the scientific communication system has been called into question. Documented abuses, such as honorific authorship, have serious implications in terms of the acknowledgment of authority, allocation of credit, and assigning of accountability. Within the biomedical world, it has been proposed that authors be replaced by lists of contributors (what I shall term the radical model), whose specific inputs to a given study would be recorded unambiguously. Here, I consider the wider implications of the hyperauthorship (Cronin 2002) phenomenon for scholarly publication, as traditionally conceived.

In recent years, scores of surveys, studies, and opinion pieces have been published in the expansive biomedical literature on the subject of multiple authorship (King 2000). While coauthorship and collaboration studies have been carried out in many disciplines and fields (Bordons and Gómez 2000), the scale of the phenomenon and associated ethical abuses have proved to be singularly problematic in the biomedical domain (Houston and Moher 1996). The central issue is not just one of multiple authorship but of hyperauthorship—massive levels of coauthorship. Some consequences of these trends have been extensive editorial commentary and correspondence in the leading biomedical journals, revised editorial guidelines for authors and collaborators submitting to reputable journals, and, most radically, a proposal to replace authors entirely with lists of contributors (Rennie, Yank, and Emanuel 1997). Curiously, the perceived seriousness of the problem does not find echo in other scientific fields. This chapter (a) begins with a brief, historical overview of scholarly publishing, focusing on the role of the author and the constitution of trust in scientific communication; (b) offers an impressionistic survey and analysis of developments in the biomedical literature; (c) explores the extent to which deviant publishing practices in biomedical publishing are a function of the sociocognitive and structural characteristics of the discipline by comparing biomedicine with high-energy physics, the only other field that appears to exhibit comparable hyperauthorship tendencies; and, finally, (d) assesses the extent to which current trends in biomedical communication may be a harbinger of reform in other disciplines.

A (VERY) BRIEF HISTORY OF AUTHORSHIP

According to Manguel (1997, 182–83), the earliest named author in history was the Mesopotamian Princess Enheduanna, who, more than 4,000 years ago, signed her name at the end of tablets, on which were etched songs in honor of Inanna, goddess of love and war. Over the centuries, however, certain genres of text (e.g., epic poems and sagas) did not always require authors: "Their anonymity was ignored because their real or supposed age was a sufficient guarantee of their authenticity" (Foucault 1977, 125). For Barthes (1977, 142–43), though, the author really is "a modern figure . . . emerging from the Middle Ages with English empiricism, French rationalism and the personal faith of the Reformation." While critical theorists may question the "prestige of authorship" and "all manifestations of *author*-ity" (Birkerts 1994, 158–59)—which helps explain the predilection for postmodernist titles such as *The Death of the Author* (Barthes 1977), *What Is an Author?* (Foucault 1977), and *The Death of Literature* (Kernan 1990)—there is little doubt that both the symbolic and material consequences of authorship today are rather more far-reaching than in Enheduanna's time. To state the obvious, public affirmation of authorship is absolutely central to the operation of the academic reward system, whether one is a classicist, sociologist, or experimental physicist (Bourdieu 1991; Cronin 1984; Franck 1999). But while the traditional or Romantic notion of authorship persists, most noticeably in the humanities, it is no longer the sole or dominant model in certain scientific specialties.

Before the precursors of today's scholarly journals had established themselves in the second half of the seventeenth century, scientists communicated via letters: "The modern scientific article . . . began life discretely, as the lecture and the letter" Montgomery (2003, 78). Letter writing was the principal means for exchanging ideas and experimental results. Gentleman scholars, including such prodigious correspondents as Samuel Hartlib, intelligencer extraordinaire; Henry Oldenberg, secretary of the Royal Society in London; or Friar Mersenne in France were the scientific community's prototypical gatekeepers, orchestrating the flow and documentation of ideas on behalf of their peers across Europe (Knight 1976; Rayward 1992). The establishment, in France in 1665, of *Le Journal des Sçavans* and, some months later, in London, of the *Philosophical Transactions* of the Royal Society (or, as it was originally called, the *Philosophical Transactions: giving some Accompt of the present Undertakings, Studies and labours of the Ingenious in many considerable parts of the World*) constituted the beginnings of the journal-based, scholarly communication system as we know it today—and the advent of "the modern authorial persona" (Johns 2003, 67). Nonetheless, letter writing continued as a medium for the informal exchange of information and for requesting fellow scientists to replicate experiments (Manten 1980, 8).

Despite the acknowledged importance of these two contemporaneous publishing developments, much has changed in terms of the ways, both instrumental and stylistic, in which scientists communicate the results of their research to their peer communities, though, in the intervening 300-plus years, certain symbolic and rhetorical practices, notably the assertion and defense of authorship, and all the presumptive rights associated therewith, have remained center stage. In the seventeenth century, the business of authorship, as the business of science itself, was much less complicated and contentious than today—which is not to say that priority disputes were unheard of, that egos were never bruised, or that "the bauble fame" did not come into play in earlier times. The following excerpt from one of Charles Darwin's (admittedly much later) letters makes touchingly clear the recurrent tension between intrinsic and extrinsic motivation, which he, and countless other scientists, both then and now, have experienced in the course of their careers: "I am got most deeply interested in my subject; though I wish I could set less value on the bauble fame, either present or posthumous, than I do, but not to any extreme degree; yet, if I know myself, I would work just as hard, though with less gusto, if I knew that my Book wd be published for ever anonymously" (Darwin 1919, 452).

Authorial rights were not always asserted with the kind of forcefulness we take for granted today—think of the often heated disputes that routinely flare up over the ordering, inclusion, and elision of names attached to multiauthored papers (Wilcox 1998). As Rennie and Flanagin (1994) remind us, there are neither standard methods for determining order nor universalistic criteria for conferring authorship status; practices vary greatly and confusingly, both within and across disciplines, ranging from alphabetization through weighted listing to reverse seniority (Riesenberg and Lundberg 1990; Spiegel and Keith-Spiegel 1970). However, much has changed since the seventeenth century, when there was no established convention for the naming of authors of scientific communications. As Katzen (1980, 191) notes in her analysis of early volumes of the *Philosophical Transactions*: "no attempt is made to give prominence to the author of the article . . . there is generally no reference at all to the author in the heading that signals a new communication. If the author is referred to in the title, it is likely to be in an oblique form . . . we are at the threshold between anonymous and eponymous authorship." Foucault (1977, 126), in fact, talks of scientific texts "accepted on their own merits and positioned within an anonymous and coherent conceptual system of established truths and methods of verification." This, as Shapin (1995, 178) notes in his brilliant exposition of the role of trust in seventeenth-century English science, was in keeping with prevailing notions of civility and gentlemanly conduct in society at large: "reluctant authorship . . . was a standard trope of early modern culture," albeit one that withered swiftly in the face of plagiaristic practices. Inconceivable though the idea of anonymous authorship may be in today's highly competitive publishing marketplace, where reputation, career success, and, ultimately, remuneration are tightly coupled with publication salience and subsequent citation, the importance of claim staking and priority determination were really only beginning to be perceived as critical issues in the formative world of seventeenth-century scientific communication.

The standard model of scholarly publishing assumes a work written by *an* author. Typically, a single author receives full credit for the opus in question. By the same token, the named author is held accountable for all claims made in the text, excluding those attributed to others via citations. The appropriation of credit and allocation of responsibility thus go hand-in-hand, which makes for fairly straightforward social accounting. The ethically informed, lone scholar has long been a popular figure, in both fact and scholarly mythology. Historically, authorship has been viewed as a solitary profession, such that "when we picture writing we see a solitary writer" (Brodkey 1987, 55). But that model, as Price (1963) recognized more than four decades ago, is anachronistic as far as the great majority of contemporary scientific, much social scientific, and even some humanistic publishing is concerned. In general terms, such stereotyping ignores the fact that a great deal of the scholarly literature is the product of a "socio-technical production and communications network" (Kling et al. 2000), which brings together a mix of actors, resources, tools, and rules. To some extent, authorship has become a collective activity, with numerous coauthors competing for the byline (Kassirer and Angell 1991; McDonald 1995). Parallel, though not quite so dramatic, growth has been observed in the number of individuals being formally acknowledged in scholarly journals for their multifarious contributions, what is sometimes referred to as subauthorship collaboration (Cronin 1995, 2001; Heffner 1979, 1981; Patel 1973)—an important indicator of informal scientific collaboration.

COLLABORATION AND THE CONDITIONS OF WORK

If the seventeenth century was a watershed in the history of scholarly publishing, the same may also reasonably be said of the mid-twentieth century. After World War II, collaboration became a defining feature of "big science" (Bordons and Gómez 2000; Cronin 1995, 4–13; Katz and Martin 1997). Major scientific challenges (splitting the atom, putting a man on the moon, mapping the human genome) typically require enormous levels of funding to support the costs of industrial-scale equipment and instrumentation, as well as complex (transnational) teams of researchers drawn from multiple disciplines and institutions. In some domains, pathbreaking work is necessarily the outcome of collaborative activity rather than individual scholarship, a fact reflected in the modest proportion of federal research funds which is allocated to individual investigators rather than teams. Collaborations are a necessary feature of much, though by no means all, contemporary scientific research.

The nature and conduct of scientific research have changed enormously since the seventeenth century. One well-documented corollary of interinstitutional, intersectoral, and international collaboration has been the striking increase in rates of coauthorship, though the latter is only a partial indicator of the former; coauthorship and collaboration are not, by the way, coextensive (Katz and Martin 1997, 1). This trend is most noticeable in experimental high-energy physics (HEP), with its often very large teams and highly sophisticated collaborations (Kling and McKim 2002). A similar trend, dating from the 1990s, can be seen in the biomedical research literature, particularly with regard to publications arising from large, multi-institutional clinical trials (Horton 1998; Rennie, Yank, and Emanuel 1997). However, it would be erroneous to conclude that collaboration and coauthorship are exclusively late-twentieth-century phenomena. The growth of scientific collaboration, as reflected in coauthorship, is directly linked to the professionalization of science. Beaver and Rosen (1978) have shown how the differential rates of scientific institutionalization in France, England, and Germany are mirrored in the relative output of coauthored papers. At the end of the eighteenth and beginning of the nineteenth centuries, French science was much more professionalized and institutionalized than was the case in either of the other European powers. Specifically, Beaver and Rosen found that more than half of all the coauthored scientific articles in their historic sample were produced by French scientists. Their study is a useful corrective to the assumption that collaboration and coauthorship were unheard of prior to the emergence of big science. In a sample of 2,101 scientific papers published between 1665 and 1800, they found that just over 2 percent described collaborative work. Notable was the degree of joint authorship in astronomy, particularly in situations where scientists were dependent on observational data.

Since the early days of professional science, there has been, in extreme cases, an order of magnitude increase in coauthorship levels, with the most dramatic growth occurring in the last couple of decades. The trend, though most visible in science, has been documented in many fields and disciplines, which is not to say that sole authorship, both of journal articles and monographs, is not still common practice in fields such as women's studies (Cronin, Davenport, and Martinson 1997). By way of example, Endersby (1996) has analyzed trends in, and reasons for, collaboration and multiple authorship in the social sciences. Patel (1973) has described the growth of coauthorship in sociological journals for the period 1895 to 1965. Bird (1997) has found evidence of coauthorship growth in the literature of marine mammal science for the period 1985 to 1993, while Koehler et al. (1999) established that the average number of authors per article in the *Journal of the American Society for Information Science* (previously *American Documentation*) rose from approximately 1.2 in the 1950s to 1.8 in the 1990s.

Data produced by the Institute for Scientific Information (ISI) show, inter alia, that the number of papers with 100 authors—a leading indicator of hyperauthorship?—increased from 1 in 1981 to 182 in 1994 (McDonald 1995) and that the average number of authors per paper in the *Science Citation Index* (*SCI*) increased from 1.83 in 1955 to 3.9 in 1999 (Atkins 2000). To take a couple of random examples, a 1997 article in *Nature* (cited almost 600 times since then) on the genome sequence of a bacterium has 151 coauthors, drawn from dozens of research laboratories scattered across twelve countries (Kunst et al. 1997). A two-page article (Daily et al. 2000) in *Science* on the economic value of ecosystems has no fewer than seventeen authors and five acknowledgees. Illustrative of the trend in biomedicine is the appreciable change exhibited by the *New England Journal of Medicine* (*NEJM*). A hundred years ago, 98 percent of the articles published in the (precursor of the) *NEJM* were sole authored; today, the figure is less than 5 percent, with the mean number of authors per *NEJM* article being six (Constantian 1999). A study of all clinical, anatomic, and laboratory investigations reported in the *Journal of Neurosurgery* and *Neurosurgery* between 1945 and 1995 found that the average number of authors per scientific article rose from 1.8 to 4.6 (King 2000). In many scientific fields, multiple rather than single authorship is now the norm (Harsanyi 1993), which brings to mind Newman's (1996, 1) apt phrase, "writing together separately." What is not entirely clear is the extent to which, if at all, publication genre affects rates of coauthorship. In other words, will the picture vary depending on whether the unit of analysis is clinical reports, scientific articles, or review papers? I return to this issue later. In short, collaborative research and publishing practices in the late twentieth century have created challenges for established notions of authorship. Ineluctably, the recognizable voice of the individual author is being replaced by the sometimes pasteurized prose of the collaboration, reinforcing the already well-established "conventions of impersonality" in scientific writing (Hyland 1999, 355).

Long before the modern practice of anonymous (double-blind) peer review became an established component of the postwar scientific bureaucracy (Chubin and Hackett 1990, 19–24), experimental claims were subjected to personal witness and supported by sworn testimony (Gross, Harmon, and Reidy 2000) either by members of scientific societies or others deemed to be of appropriate social standing—a process that Shapin (1995, xxi) refers to as "the gentlemanly constitution of scientific truth." Bazerman (1988, 140), in describing the process of making witness in the early days of the Royal Society, notes that one rhetorical strategy is "to establish ethos," that is, convey to a critical audience that the author or observer of the experiment being described is a credible witness. Just as Princess Enheduanna "enabled the reader to read a text in a given voice" by signing her name to her songs (Manguel 1997, 182), so might Fellows of the Royal Society seek to create a culturally acceptable authorial persona, or ethos, to engage and persuade their audience. Or, as Montgomery (2003, 13) puts it, "The witnessing 'I' was thus science's first storyteller. It was a way to 'prove' rhetorically that the work had actually been done." But circumstances have changed such that today the individual author's voice is almost imperceptible. It is difficult to attribute a persona to an ensemble of putative authors/collaborators, or even imagine how such a confederacy might establish an ethos, though such lengthy lists of authors, doubtless, add to the perceived credibility and authority of the text. Mikhail Epstein (2000), a cultural theorist, uses the term "hyper-authorship" to capture the sense of authorship "dispersed among several virtual personalities which cannot be reduced to a single 'real' personality."

In some domains, individualistic notions of authorship ("the fiction, the voice of a single person" [Barthes 1977, 143]) have given way to the idea of the author as a collective, whose members are often geographically distributed, and, not infrequently, drawn from quite different professions and specialties. Duncan (1999), in fact, uses the analogy of a partnership with contractual ties to capture the complexities of the contemporary research and publication system. Another analogy, from the worlds of painting, where "much of eminent art is collective" (Steiner 2003, 132), and also architecture, might be the (virtual) atelier, with a team of proto-authors working under the direction of the master crafts-man, or super PI (principal investigator). This is not entirely without historical precedent; Shapin's (1995, 367) detailed account of Robert Boyle's scientific milieu includes a compelling image of "the seven-teenth-century laboratory as collective workshop." As a result, the au-thor's persona, insofar as the impersonal and highly stylized format of the scientific article permits the extrusion of an authorial self, has been dislodged by the depersonalized voice of the ensemble. Such practices, and associated concerns about "unearned authority" (Kitcher 1993, 315) and dilution of responsibility, have provoked radical proposals, no-tably in the biomedical research community—where multiple author-ship has metastasized into hyperauthorship—to restore authorial ac-countability and rationalize the allocation of academic credits. Some medical commentators (Rennie, Yank, and Emanuel 1997, 582) have advocated abandoning the concept of author altogether in favor of "con-tributors" and "guarantors," thereby freeing us "from the historical and emotional connotations of authorship." The potential subversion of the historical conception of authorship, with, at the risk of oversimplifying, its assumption of singularity, by contemporary scientific practice has created a number of ethical and procedural difficulties for those re-sponsible for evaluating the nature of individual contributions, such as university promotion and tenure committees, and also for those con-cerned with quality assurance in complex, multi-institutional research projects. In biomedicine, the standard model (one text, one author, in its primitive manifestation) faces a determined challenge from the radical model (lists of contributors and their specific inputs), championed most vigorously in recent years by Drummond Rennie, a deputy editor of the *Journal of the American Medical Association* (*JAMA*).

ASSUMING AUTHORSHIP

The central issue can be framed in terms of Rennie and Flanagin's (1994, 469) beguilingly simple question: "[H]ow many people can wield one pen?" (It is worth noting, in passing, that catalogers have traditionally dealt with this issue—"diffuse authorship"—by invoking the so-called rule-of-three: when a work has more than three authors, only the first will be listed.) The sovereignty of the standard model is being hotly contested in biomedical research fields where intense levels of professional collaboration and coauthorship are commonplace (Croll 1984; King 2000; Rennie and Flanagin 1994; Rennie and Yank 1998; Rennie, Yank, and Emanuel 1997). Proposals for reform that seek to retire the concept of authorship and replace it with a scheme for the allocation of specific, task- or job-related credits (Smith 1997; Squires 1996) are not only being debated by editors and others but are being adopted by some of the leading scientific journals. To understand why such radical thinking would attract serious attention in the medical establishment, we need to consider how multiple authorship, in extremis—what I have chosen to term "hyperauthorship"—undermines commonly held assumptions about the nature and ethical entailments of authorship, and how, in exceptional cases, it can lead to fundamental questions about the integrity of the research community as a whole. Unfortunately, little effort is made in the biomedical literature to distinguish systematically between what might casually be termed acceptable levels of multiple authorship and unacceptable levels of hyperauthorship. Studies of coauthorship trends are inconsistent in defining and operationalizing the phenomena under consideration, such that discussion of coauthorship easily leads into condemnation of hyperauthorship.

Under the standard model, the rights and responsibilities of authorship are clearly apprehended by all parties—authors, editors, referees, and readers. In attaching my name to this book, I am nailing my flag to the mast; if the book attracts critical approval, is discussed, quoted and, in due course, cited in the scholarly literature, I shall be happy to bank the symbolic capital that accrues to me as author and originator. If it is challenged because of exiguity of theoretical, historical, or empirical heft, I shall simply have to face the music: there are no coauthors to help deflect the charge of sciolism. Likewise, if I am challenged for drawing too sparingly, selectively, or indulgently on the ideas and work of others, I understand the possible consequences. However, I have chosen neither to hide behind the cloak of anonymity nor to bypass the rigors of editorial review by posting the manuscript on my website; rather, I want to attract the attention of my peers while signifying my trustworthiness, and the best way still to do that is to publish formally. As a serial author, I am fully cognizant of the rights and responsibilities of authorship. I understand the norms of scholarly publishing, and I am aware of the sanctions that may be invoked if infractions occur. Should the arguments in *The Hand of Science* prove flawed, no one but myself is to blame, and that includes those whom I have named in the acknowledgments. If the book attracts attention, I shall be happy to bask in the glow. Just as would a lawyer, doctor, or dentist, I implicitly accept full responsibility for the work to which I affix my name and, by extension, my professional reputation.

Unquestionably, the practice of promiscuous coauthorship puts considerable stress on this tried and tested model. In the case of a paper with, say, two or three coauthors, one of whom is the designated corresponding author (to whom editors and readers channel their queries, comments, and concerns), this should not be a serious problem, as the specific contributions of each individual presumably can be clearly described to the satisfaction of all parties. However, if the number of listed coworkers is counted in dozens, scores, or even hundreds—as is increasingly common, for instance, in the case of large-scale clinical trials—who, then, is to be held accountable for both the whole and its constituent parts, should questions about the integrity of the work arise? And who, remote though the possibility may be, will act as spokesperson in the face of a liability suit brought against the (multidisciplinary) collaboration? Likewise, how should credit for crafting a single publication be allocated across teams comprising dozens or scores of coworkers whose membership may have altered over the course of a multiyear project? Where, ultimately, do authority, credit, and accountability reside? In the words of Rennie and Yank (1998, 829), when "the number of collaborators grows arithmetically, it becomes exponentially harder to affix responsibility." At this point, the notion of authorship, literally interpreted, is effectively rendered meaningless.

Apart from biomedicine, these issues have also attracted attention in the high-energy physics research community, where hyperauthorship is commonplace. However, multiple authorship and hyperauthorship are not problematized by physicists as they are by the biomedical community. Common to both domains, though, has been discussion of how best to apportion credit when there are many coauthors (McDonald 1995). How, for instance, should a promotion and tenure committee view the contribution of the ninety-ninth listed author on a particle physics paper or the thirty-sixth author on a genome sequencing study? What may seem to constitute a miniscule portion of a single journal article may, in fact, have consumed a significant amount of that individual's professional time and energy. Two other issues, honorific authorship and data integrity, seem to be of especial concern to the biomedical community, given widespread media coverage of, and speculation about, fraudulent practice, the effects of which, in both career and personal terms, can be devastating (Kevles 1998). What this seems to suggest is that size of collaboration alone does not necessarily engender concerns about the veracity and reliability of the reported research. It is likely that the sociocognitive structures of these two research cultures, and their relative transparency, differ in important respects. As a result, issues of social trust are manifestly more problematic in the discourse of the biomedical community. I return to the matter of field dependence and generalizability later.

MULTIPLE AUTHORSHIP IN BIOMEDICINE

Many studies have documented the recent dramatic increase in multiple authorship in medical and related fields, in turn generating a mass of editorial commentary and correspondence in the letters pages of major journals (Constantian 1999; Rennie, Yank, and Emanuel 1997). However, as noted earlier, it is not always easy to establish what constitutes unacceptable levels of multiple authorship in the minds of the many concerned commentators, be they authors or editors. Individual research and sole author publication are almost inconceivable in certain biomedical specialties, where complex (national and international) networks of resources, facilities, technical expertise, and subjects are necessary ingredients of serious, funded research. Collaboration and coauthorship are inescapable consequences of increasing specialization and technical sophistication. In short, structural interdependence is a fact of life in the age of industrialized medical research, and the phenomenon is powerfully revealed in the lengthy lists of authors and acknowledgees that adorn the majority of published articles and reports. As far as biomedical research is concerned, it seems that the lone scholar/independent researcher model is a thing of the past. To be sure, coauthorship in itself is not inherently problematic, but some of the hidden social practices which it has occasioned, such as receiving credit under false pretenses, are cause for grave concern (Anderson 1991). Even a cursory examination of the mainstream biomedical literature over the last decade or two makes this fact abundantly clear. In particular, the abuses implied by the term "honorific authorship" are damaging the perceived credibility and integrity of the medical communication system as a whole, a development that has been exacerbated by the number of high-profile cases of actual or alleged scientific fraud in recent years (Cohen 1999; Eysenck 1999; Relman 1983). Horton (1998, 688), editor of *The Lancet*, speaks, in fact, of "the shattered system of academic reward and its symptom, broken rules of authorship," a view which seems neither extreme nor marginal judging by the tenor of the debate being conducted in the pages of the biomedical literature (Klein and Moser-Veillon 1999).

Although the increase in coauthorship is demonstrably a function of scale and specialization effects in medical and related research, the picture is more complicated still. There is persuasive evidence that many individuals receive unwarranted coauthor status (variously referred to as "guest," "gift," or "surprise" authorship) while others are denied legitimately earned author status ("ghost" authorship). As Slone (1996, 578) notes, authorship "cannot be conferred but must be earned." In a survey of three large- and three smaller-circulation medical journals (*Annals of Internal Medicine*, *Journal of the American Medical Association*, *New England Journal of Medicine*, and *American Journal of Cardiology*, *American Journal of Medicine*, *American Journal of Obstetrics and Gynecology*, respectively), Flanagin et al. (1998) developed a multivariate logistic regression model to test the hypothesis that coauthored articles (operationalized as papers with six or more authors) were increasing at a rate greater than would be expected when confounding variables, such as the number of centers, were taken into account. They found that 19 percent of original research reports had honorific authors. In other words, almost one in five authors in leading medical journals were garnering unwarranted credit and phantom fodder for their curricula vitae. They also discovered that 1 percent of articles had ghost authors, which means that quite a few individuals were not receiving due credit for their creative or material contributions to the research process—those whom Shapin (1995, 379) describes as "the ghostly inferred hosts of unnamed actors who shifted instruments about and exerted their muscular labor."

Flanagin et al.'s (1998) findings, based on surveys of corresponding authors, are in keeping with other estimates of honorific authorship in the biomedical literature.

Slone (1996), in a survey of "major research" articles published in the *American Journal of Roentgenology*, found that undeserved coauthorship rose from 9 percent on papers with three coauthors to 30 percent on papers with more than six coauthors. Drenth (1998) found that the mean number of authors per article in the *British Medical Journal* rose from 3.2 in 1975 to 4.7 in 1995. A number of publication genres featured in these and other studies. For example, King (2000) examined "research articles"; Kahn et al. (1999) examined two kinds of "articles," "randomized controlled trials" and "controlled observational studies"; Slone (1996) restricted his focus to "major papers"; Drenth (1998) examined "original articles"; and Flanagin et al. (1998) investigated "research articles," "review articles," and "editorials." It is not clear from the various analyses whether, or to what extent, genre and coauthorship levels are linked. Further confirmation that inflationary trends in authorship are not explicable purely in terms of intercenter collaboration, research funding, and related factors has been provided by Kahn et al. (1999). Additionally, Ducor (2000) investigated a small set of patents in molecular biology and their concomitant publications in the scientific literature. Of the forty patent-article pairs examined, all but two listed more authors than inventors, which raises interesting questions about the relative stringency of the criteria employed for conferring authorship and inventorship.

Such studies reveal the apparent ease with which authorial spurs can be earned. However, as Slone (1996, 576) points out, there is what might, at first blush, appear to be a fairly simple quid pro quo in operation—bylines for bodies: "The number of coauthors in multiinstitutional clinical trials is understandable, and it is tempting in complex studies to offer authorship to referring physicians and ancillary medical specialists to obtain the needed cooperation." These and kindred practices may well help explain the startlingly high and sustained rate of scholarly production demonstrated by some biomedical researchers (Anderson 1992). Whether coat-tailing on one's peers and rent-taking are dismissed as understandable peccadilloes or seen as a corroding of the norms and institutional values that undergird the social system of science (Merton 1973), the practical consequences are ramified in cases of fraud. In such instances, readers and editors may find that nominal authors adroitly disclaim responsibility and proffer a variety of excuses to ensure that culpability falls elsewhere. In rare cases, a questionable published paper may acquire "orphan" status (Rennie and Flanagin 1994, 469), as all concerned try to wash their hands of it, invoking hyper-labor specialization as grounds for exoneration. Such a scenario is inconceivable under the standard model, where authorship and accountability are isomorphic. But when authorship/ownership of a study is distributed across multiple contributors, many of whom may have zero or weak relationships, whether personal or institutional, with their myriad coworkers (Katz and Martin 1997), the practical (i.e., enforceable) allocation of accountability may pose intractable problems.

AUTHORSHIP AND ACKNOWLEDGMENT

Historically, authorship implied writing. Today, the concept of authorship has expanded to accommodate a diverse array of contributions and inputs, some of which may require little or no engagement with the text qua text. This broadening of definition finds its echo in the world of cataloging and metadata, where schemes such as the Dublin Core and REACH (Record Export for Art and Cultural Heritage) prefer the terms "creator" and "maker" to the more traditional "author" (Baca 1998). The Dublin Core Metadata Initiative (see http://purl.org/dc/documents/rec-dces-19990702.htm) includes the element "Contributor" as someone who is defined as being "responsible for making contributions to the content of the resource." In addition, there is the element "Other Contributor," which can accommodate a "person or organization not specified in a Creator element who has made significant intellectual contributions to the resource but whose contribution is secondary." The new inclusiveness defines authorship of a scientific article as requiring "expertise in its content and thorough knowledge of the investigation reported" (Gaeta 1999, 297). Under this not unreasonable-sounding definition, the author need not write a word. Within the biomedical community, considerable effort has been invested in developing guidelines and criteria for authorship, notably those adopted by the International Committee of Medical Journal Editors (ICMJE 1997), also known as the Vancouver Group. But before considering the ICMJE's guidelines for authorship, it may be helpful to reflect briefly on the role of acknowledgments in the primary communication process, and the relationship between authorship and acknowledgment.

The explosion of coauthorship naturally raises the question why authorial surplus could not be accommodated in the acknowledgments that accompany the great majority of scientific articles. The acknowledgment serves as a parking lot for miscellaneous contributions—cognitive, technical, and social—that typically fall short of meeting commonly understood criteria for awarding coauthorship. But we should not assume too much in terms of common understandings; the dividing line between the two classifications, author and acknowledgee, is neither universally appreciated nor consistently applied. I have shown (Cronin 1995, 85–86) that interpretative disputes are not uncommon, and that some researchers feel that they have been denied their just deserts by being downgraded from coauthor to acknowledgee. Ambiguities and fuzzy interpretations notwithstanding, the ubiquity and significance of "sub-authorship collaboration" (Patel 1973, 81), otherwise conceptualized as "trusted assessorship" (Mullins 1973, 32), has been amply documented in recent years. Bazerman (1984, 1988) has chronicled the evolution of the acknowledgment during the nineteenth and twentieth centuries in the journal literature of experimental physics, showing how it became, to paraphrase Grafton (1997, 233), an integral part of the rhetoric of narration and annotation. Others have recorded the social significance of acknowledgment practices in a variety of disciplines, including astronomy (Verner 1993), genetics (McCain 1991), biology (Heffner 1981; Laband and Tollison 2002), chemistry (Heffner 1981), psychology (Cronin 1995; Heffner 1981), information science (Cronin 1995), sociology (Cronin 1995; Patel 1973), political science (Heffner 1981), economics (Laband and Tollison 2000), and philosophy (Cronin 1995). It is abundantly clear from the findings of these studies that acknowledgment is a well-established and valuable feature of the apparatus of twentieth-century scholarly communication, one that can shed additional light on "institutional dependencies" (Bazerman 1984, 183) not fully reflected in the article's byline.

At first glance, the idea of using acknowledgments as a form of triage, leading to "two-tiered authorship" (Saffran 1989, 9), would seem to make sense, assuming, for the sake of argument, that consensus could be reached on criteria for allocating coworkers a byline as opposed to "mere" acknowledgment status. But things are not quite so simple.

More than a decade ago, Kassirer and Angell (1991, 1511) of the *New England Journal of Medicine* were bemoaning not only "ambiguous authorship" but also "lengthy acknowledgments," inflated by the inclusion of "everybody who had anything to do with the study, including those who were merely carrying out their jobs, such as technicians." Their radical proposal at that time was to limit the amount of journal space devoted to acknowledgments, with the excess names being deposited with the National Auxiliary Publications Service—scholarly publishing's equivalent of the *salon des refusés*. Some journals place a limit on the number of coauthors; for example, the *Dutch Journal of Medicine* does not publish articles with more than six authors (Hoen, Walvoort, and Overbeke 1998). The problem with arbitrary capping, whether of authors or acknowledgees of one kind or another, is that some individuals' potentially important contributions, be they clinical investigators (Carbone 1992) or telescope operators (Cronin 1995), may be erased. This could, conceivably, have negative downstream implications in terms of remuneration and promotion prospects for those—the "invisible technicians" (Shapin 1995, 355)—whose efforts have been withheld from the public ledger. It might also reduce, in line with theories of reciprocal altruism (Nowak and Sigmund 2000), potential collaborators' willingness to "donate" their services.

Overpopulation is now a feature of both author lists and acknowledgments, and, as already stated, the picture is muddied by the lack of widely accepted guidelines for allocating credits. Confusion over the criteria for awarding authorship as opposed to acknowledging colleagues—and acknowledgment, it should be noted, is often subdivided into intellectual and technical, which further complicates the calculus—has convinced Rennie, York, and Emanuel (1997) that the distinction between the two modes of credit allocation is inherently artificial. Consequently, they have argued for explicit description of *all* individual contributions as a means of eliminating ambiguity. Such a proposal would remove both authorship and acknowledgment from the frame, a really quite significant break with scholarly publishing tradition.

ASSESSING AUTHORSHIP

More than 500 journals have adopted the ICMJE's (1977) principles of authorship as laid out in the fifth edition of the Uniform Requirements for Manuscripts Submitted to Biomedical Journals (Klein 1999; Stern 2000). According to these guidelines, candidate authors must satisfy three conditions. They must make "substantial contributions to (a) conception and design, or analysis and interpretation of data; and to (b) drafting the article or revising it critically for important intellectual content; and on (c) final approval of the version to be published." Laudable though these guidelines are, it is unlikely that they can solve the problem. The study by Hoen et al. (1998) in the Netherlands found that authors and their coauthors did not always agree with one another's assessments that the ICMJE criteria had been met. In the United Kingdom, Bhopal et al.'s (1997) survey of medical researchers discovered that, although most respondents concurred with the three criteria (more than 80 percent in each case), a majority (62 percent) did not feel that all three conditions should have to be satisfied to warrant author status. Further, more than half the respondents believed that the ICMJE's criteria were not usually adhered to by their peers. What these studies suggest is that even the best-intentioned guidelines for authors, such as those specified by the *Journal of the American Medical Association* (see http://jama.ama-assn.org/info/auinst.html), will be variously and inconsistently interpreted. Absent unrealistic, and presumably unacceptable, levels of policing by editors and their advisory boards, it seems unlikely that illicit authorship and cronyism will be extirpated as a result of such guidelines.

The standard model accepts that authorship is linked inextricably to writing. But writing is no longer a necessary condition of coauthorship in certain cases. Thus, an alternative to authorship is required to accommodate the many other contributions that shape the published byproducts of collaborative activity, be they research reports, journal articles, conference papers, or technical reports. Comparisons have been made (Rennie, Yank, and Emanuel 1997) with the credit sequences at the end of movies, in which all those responsible, from key grip through stunt men to director, are thanked publicly for their services. Indeed, I have analogized in similar fashion, entitling an early paper on acknowledgments, "Let the Credits Roll . . ." (Cronin 1991). However, it is important to distinguish between generic job categories and the specification of tasks performed; the contributorship model is designed to record each individual's actual input (e.g., experimental design, data collection, statistical analysis, final article revision), not job title (e.g., co-principal investigator, technician, systems analyst), since the latter may on occasion mask or inflate the former (Stern 2000).

The general case for replacing authors with lists of contributors and guarantors has been most cogently detailed by Rennie, Yank, and Emanuel (1997) in the pages of *JAMA*. Their paper systematically addressed the pros and cons of the radical model. In this scheme of things, project coworkers would discuss and agree upon the distribution of the various responsibilities and tasks at the outset, assess the relative merit of their contributions, and determine the implications for listing collaborators' names, so that first authorship would be based on earned credit rather than alphabetic advantage or professional status. Appropriately, the final paragraph of their *JAMA* manifesto is itself a detailed itemizing of the specific contributions of each coauthor, such that the reader is left in no doubt as to who did precisely what. The great attractions of the scheme are the implications for transparency and equity, but there is also no doubt that serious-minded application of the contributorship model would impose additional overhead on medical researchers, and team leaders in particular. This would certainly be true if individual contributions in a large multicenter trial were to be calibrated with the degree of meaningful precision described by Ahmed et al. (1997).

Their authorship matrix (which they apply in exemplary fashion to their own article) combines key tasks associated with multiauthored papers (e.g., conception, design, data analysis), with each investigator's contribution(s) duly ranked on a three-point scale (minimal, some, significant). The respective weights (1, 3, and 5) are summed to (a) estimate each individual's overall contribution, which can range from 1 to 35, and (b) determine the sequence in which coauthors' names are listed on the resultant publication. The logic behind the algorithmic approach is impeccable, but its broad practicability remains unproven. Other, not entirely unrelated, efforts to increase transparency in the scientific communication system have proved unsuccessful. For instance, it was suggested in the 1980s that authors should classify the citations attached to their papers in terms of the functions they performed, but the idea failed to get off the ground, despite some early interest from the Institute for Scientific Information in Philadelphia (Cronin 1984; Duncan, Anderson, and McAleese 1981). In the early 1990s, skepticism also greeted the proposal that an online acknowledgment index, akin to the *Science*

Citation Index, be developed to record subauthorship collaboration in science (Cronin and Weaver-Wozniak 1992).

While listing contributions may clarify the nature of coworkers' participation and, thus, both reduce the incidence of honorific authorship and ensure more equitable allocation of credit, it does not necessarily address the thorny issue of ultimate responsibility for the overall integrity of the study. Consequently, Rennie, Yank, and Emanuel (1997) have also proposed that special contributors be designated who would act as owners, or guarantors, of the work as a whole. These individuals would be expected to ensure the integrity of the work and be able to defend the research team's findings in the event of peer challenge or public criticism. In the matrix adumbrated above (Ahmed et al. 1997, 43), the dimension "public responsibility" is included to signify that an author participated in, and is accountable for, the work and related results. Despite continuing discussion of the merits and shortcomings of the radical model (e.g., Should acknowledgments and contributor lists be combined? Is it reasonable to expect any individual to be able to account for all aspects of a complex multidisciplinary collaboration?), some editors and their journals (e.g., *The Lancet*, *British Medical Journal*) are supportive of the idea of contributor lists (Horton 1998).

Are the developments in biomedical research and publication sketched here peculiar to that domain or do they occur elsewhere? To put it another way, and to paraphrase Mark Twain, are the rumors of the death of authorship greatly exaggerated? Although increasing rates of multiple authorship have attracted attention in a number of fields, such as psychology (Bartle, Fink, and Hayes 2000), only in biomedicine has the research community felt a need for collective, aggressive action to deal with the excesses and documented abuses. Why, then, does the same not hold true for high-energy physics, the other acknowledged locus of hyperauthorship? In HEP, the massive scale and cost of conducting pathbreaking research necessitate resource concentration around a number of major laboratories (e.g., CERN, Fermilab, SLAC), at which large (distributed) collaborations are instituted to tackle basic research problems, typically requiring multiyear funding (Traweek 1992). Such highly complex, distributed sociotechnical ensembles (Kling et al. 2000) can bring together literally hundreds of physicists from countries, laboratories, and universities worldwide, who compete directly with collaborations centered elsewhere. These multidisciplinary teams routinely produce publications that have numerous coauthors, just as in the case of the biomedical research community. Yet, despite some concerns about proliferating HEP authorship rates, and occasional discussion of hyperauthorship in the higher education press (McDonald 1995), the kind of soul searching and public debate on authorial integrity that have characterized the biomedical community in recent years are not mirrored in the world of high-energy physics. Why should this be so?

The answer has to do with the relative intensity of professional socialization and oral communication in the two cultures (Traweek 1992, 120–23), along with the character of the organizational structures and value systems that define collaborations in large-scale, high-energy physics and biomedical research. Social coordination theory provides a potentially rich analytic framework for understanding cooperative acts. Interactants—networks of coworkers, coauthors, contributors, and acknowledgees in the present context—must satisfy a number of conditions, such as establishing copresence and revealing mutual responsiveness, for "a cooperative act to unfold" (Couch 1984; Fine 1993, 72). Ethnographic approaches could be used to explore a variety of research and publishing subcultures and document the ways in which social relationships and group behaviors mold material practice in these two domains. Second, within HEP collaborations, there is a powerful admixture of skepticism (Traweek 1992, 117) and very explicit, transparency-inducing procedures for vetting work internally. This point is stressed by Kling at al. (2000) in their study of interaction networks: "concern for professional reputation both individually and for the collaboration causes most, if not all, collaborations to have developed systems whereby only research results that have been 'blessed' by the collaboration may be shared with the world on the collaboration website." The HEP research community is thus characterized by high levels of internal scrutiny, mutual trust—witness, for instance, the institutionalized practice of relying upon, and citing, e-preprints—and peer tracking, such that it is not prone to systematic fraud. Contemporary scientific collaboration need not necessarily imply impersonality and virtual anonymity, a point made tellingly by Shapin (1995, 414–15):

> While outsiders—including many philosophers and sociologists—tend happily to refer to vast numbers of practitioners called "scientists," insiders function within specialist groups of remarkably small size. . . . The group of people mutually judged capable of participating in each of these specialized practices—what has been called the core-set—may be very small.

Impressionistically, biomedical collaborations are more heterogeneous and socially diffuse in character and do not appear to have the same degree of multilayered, internal review as HEP research collaborations. This suggests that local practices, social and organizational, shape knowledge production and dissemination behaviors in high-energy physics in ways that make it significantly different from medical and related disciplines, with which it may have surface similarities. An example would be electronic preprint archives, which are a well-established and much discussed phenomenon in the physics community but, as yet, less popular in other scientific disciplines (Kling et al. 2000). For this reason, it is wise to avoid generalizations and to concentrate instead on showing how interactions among coworkers, specifically the orchestration of information exchange and coauthorship, are grounded in local culture. That said, "the usual methods of observation, interviews or questionnaire" may lack the necessary sensitivity to capture precisely the often evanescent and inherently social nature of collaboration (Katz and Martin 1997, 2). Happily, a diverse body of work on the socially situated nature of scientific communication already exists that points the way. This ranges from Crane's (1969) pioneering analyses of invisible colleges through Latour and Woolgar's (1979) classic study of laboratory life at the Salk Institute to Traweek's (1992) richly textured ethnography of the HEP community. In addition, the work of Schatz and colleagues on the Worm Community System project, which was designed to capture the full range of knowledge, formal and informal, of the community of molecular biologists who study the nematode worm *C. elegans* (see www.canis.uiuc.edu/projects/wcs/index.html) can provide useful insights; so, too, research into the material practices and social interactions of scientists working in collaboratories, such as the Upper Atmospheric Research Collaboratory (see http://intel.si.umich.edu/crew/Research/resrch08.htm) or the Space, Physics & Aeronomy Research Collaboratory (see http://intel.si.umich.edu/sparc/) at the University of Michigan.

CONCLUSIONS

In biomedicine, authorship has irrevocably shed some of its craft associations; to be an author is not necessarily to be a writer. But that is hardly newsworthy. Contemporary science is quite different from seventeenth-century science in terms of both its social and economic structures. In many areas, interdependence is an inescapable fact of research life. Changes in the conditions of scientific work and the associated reward structures have had significant impacts on our understanding of authorship. The "modern scriptor" (Barthes 1977, 146) is no longer the sole conceiver, fabricator, and owner of the published article. Instead, today's biomedical journal article is the progeny of occasionally enormous collaborations, the individual members of which may have minimal involvement in the fashioning of the literary end product itself, with the act of writing being delegated to a subgroup or designated spokespersons. In such circumstances, the standard model, with its connotations of individualism and intimacy, seems inadequate. Ever more fine-grained division of "cognitive labor" (Kitcher 1993, 303) and ever-increasing interdependence in areas such as biomedicine and high-energy physics will require a fundamental rethinking of the concept of author.

In the contexts of biomedical and HEP research, it is difficult to reject outright the appeal of the radical model, with its bifurcation of authorship into contributorship and guarantorship, since what is implied by a byline in these cases is typically a very precise, often specialized input to a complex, multidisciplinary project. The classical idea of authorship cannot credibly accommodate the legions of coworkers associated with large-scale collaboration, nor can it adequately reflect "the epistemic role of support personnel" in the conduct of science (Shapin 1995, 359). In some, admittedly extreme, instances, authorship and writing are being decoupled. It is clear from an examination of the journal literature of biomedicine that some kind of ontological reassessment of authorship is called for to ensure that authority, credit, and accountability, currently apportioned in confused fashion across the trinity of authors, acknowledgees, and contributors, are henceforth distributed appropriately, parsimoniously, and unambiguously. It is possible that the ongoing debate in the biomedical literature will spark fresh examination of both publication and reward practices in other fields, but much more systematic investigation will be required, along with some exploration of the extent to which article genre and place of publication influence rates of coauthorship across disciplines. In the future, it is quite likely that the concept of the "author-function," to use Foucault's (1977, 124–31) term, will vary from one "epistemic community" (David 1998,

139) to the next. In some domains, the solo scholar may well take a back seat to the investigative ensemble in crafting the scientific record, but it would be grossly premature to herald the death of the author, for, in the words of Biagioli (2003, 274), "scientific authorship, whatever shape it might take in the future, will remain tied to specific disciplinary ecologies."

REFERENCES

Ahmed, S. M., Maurana, C. A., Engle, J. A., Uddin, D. E., and Glaus, K. D. (1997). A method for assigning authorship in multiauthored publications. *Family Medicine*, 29(1), 42–44.

Anderson, C. (1991). Hunger strike at volcanology institute. *Nature*, 354, 3.

Anderson, C. (1992). Writer's cramp. *Nature*, 355, 101.

Atkins, H. B. (2000). Personal communication with Helen Atkins, director of Database Development, Institute for Scientific Information, Philadelphia.

Baca, M. (Ed.). (1998). *Introduction to Metadata: Pathways to Digital Information*. Los Angeles, CA: Getty Information Institute.

Barthes, R. (1977). The death of the author. In: Barthes, R. *Image, Music, Text*. Essays selected and translated by S. Heath. New York: Hill and Wang, 142–148.

Bartle, S. A., Fink, A. A., and Hayes, B. C. (2000). Psychology of the scientist: LXXX. Attitudes regarding authorship issues in psychological publications. *Psychological Reports*, 2000, 86, 771–788.

Bazerman, C. (1984). Modern evolution of the experimental report in physics: Spectroscopic articles in *Physical Review*, 1893–1980. *Social Studies of Science*, 14, 163–196.

Bazerman, C. (1988). *Shaping Written Knowledge: The Genre and Activity of the Experimental Article in Science*. Madison, WI: University of Wisconsin Press.

Beaver, D. de B. and Rosen, R. (1978). Studies in scientific collaboration: Part I: The professional origins of scientific co-authorship. *Scientometrics*, 1, 65–84.

Bhopal, R., Rankin, J., McColl, E., Thomas, L., Kaner, E., Stacy, R., Pearson, P., Vernon, B., and Rodgers, H. (1997). The vexed question of authorship: Views of researchers in a British medical faculty. *British Medical Journal*, 314(7086), 1009–1012.

Biagioli, M. (2003). Rights or rewards? Changing frameworks of scientific authorship. In: Biagioli, M. and Galison, P. (Eds.). *Scientific Authorship: Credit and Intellectual Property in Science*. New York: Routledge, 253–279.

Bird, J. E. (1997). Authorship patterns in marine mammal science, 1985–1995. *Scientometrics*, 39(1),19–27.

Birkerts, S. (1994). *The Gutenberg Elegies: The Fate of Reading in an Electronic Age*. New York: Fawcett Columbine.

Bordons, M. and Gómez, I. (2000). Collaboration networks in science. In: Cronin, B. and Atkins, H. B. (Eds.). *The Web of Knowledge: A Festschrift in Honor of Eugene Garfield*. Medford, NJ: Information Today, 197–213.

Bourdieu, P. (1991). *Language and Symbolic Power*. Cambridge, MA: Harvard University Press.

Brodkey, L. (1987). *Academic Writing as Social Practice*. Philadelphia, PA: Temple University Press.

Carbone, P. P. (1992). On authorship and acknowledgments. *New England Journal of Medicine*, 326(16), 1084.

Chubin, D. E. and Hackett, E. J. (1990). *Peerless Science: Peer Review and U.S. Science Policy*. Albany: State University of New York Press.

Cohen, P. (1999, July 2). Your mission is . . . *New Scientist*, 38–41.

Constantian, M. B. (1999). The Gordian knot of multiple authorship. *Plastic and Reconstructive Surgery*, 103(7), 2064–2066.

Couch, C. J. (1984). Symbolic interaction and generic sociological principles. *Symbolic Interaction*, 8, 1–13.

Crane, D. (1969). Social structure in a group of scientists: A test of the "invisible college" hypothesis. *American Sociological Review*, 32(5), 335–352.

Croll, R. P. (1984). The noncontributing author: An issue of credit and responsibility. *Perspectives in Biology and Medicine*, 27(3), 401–407.

Cronin, B. (1984). *The Citation Process: The Role and Significance of Citations in Scientific Communication*. London: Taylor Graham.

Cronin, B. (1991). Let the credits roll: A preliminary examination of the role played by mentors and trusted assessors in disciplinary formation. *Journal of Documentation*, 47(3), 227–239.

Cronin, B. (1995). *The Scholar's Courtesy: The Role of Acknowledgement in the Primary Communication Process*. London: Taylor Graham.

Cronin, B. (2001). Acknowledgment trends in the research literature of information science. *Journal of Documentation*, 57(3), 423–430

Cronin, B. (2002). Hyperauthorship: A postmodern perversion or evidence of a structural shift in scholarly communication practices? *Journal of the American Society for Information Science and Technology*, 52(7), 558–569.

Cronin, B., Davenport, E., and Martinson, A. (1997). Women's studies: bibliometric and content analysis of the formative years. *Journal of Documentation*, 53(2), 121–136.

Cronin, B. and Weaver-Wozniak, S. (1992). An online acknowledgements index: rationale and feasibility. In: *Proceedings of the 16th International*

Online Meeting, London, December 8–10, 1992. Oxford: Learned Information, 281–290.

Daily, G. C. et al. (2000). The value of nature and the nature of value. *Science*, 289, July, 395–396.

Darwin, C. (1919). Autobiography. In: Darwin, F. (Ed.). *The Life and Letters of Charles Darwin*. New York: Appleton.

Drenth, J. P. H. (1998). Multiple authorship: The contribution of senior authors. *Journal of the American Medical Association*, 280, 219–221.

Ducor, P. (2000, August 11). Coauthorship and coinventorship. *Science*, 289, 873, 875.

Duncan, A. M. (1999). Authorship, dissemination of research findings, and related matters. *Applied Nursing Research*, 12(2), 101–106.

Duncan, E. B., Anderson, F. D., and McAleese, R. (1981). Qualified citation indexing: its relevance to educational technology. In: Duncan, E. B. and McAleese, R. (Eds.). *Information Retrieval in Educational Technology. Proceedings of the First Symposium on Information Retrieval in Educational Technology, 1st April 1981, Aberdeen*. University of Aberdeen, 70–79.

Endersby, J. W. (1996). Collaborative research in the social sciences: Multiple authorship and publication credit. *Social Science Quarterly*, 77(2), 375–392.

Epstein, M. (2000). Hyper-authorship: The case of Araki Yasusada. *Rhizomes*, 1. Available at www.rhizomes.net/Issue1/toc.html.

Eysenck, H. J. (1999). Why do scientists cheat? *Journal of Information Ethics*, 8(2), 27–35.

Fine, G. A. (1993). The sad demise, mysterious disappearance, and glorious triumph of symbolic interactionism, *Annual Review of Sociology*, 19, 61–87.

Flanagin, A. et al. (1998). Prevalence of articles with honorary authors and ghost authors in peer-reviewed medical journals. *Journal of the American Medical Association*, 280(3), 222–224.

Foucault, M. (1977). What is an author? In: Bouchard, D. F. (Ed.). *Language, Countermemory, Practice: Selected Essays and Interviews*. Ithaca, NY: Cornell University Press, 113–138.

Franck, G. (1999, October 1). Scientific communication—a vanity fair? *Science*, 286, 53, 55–56.

Gaeta, T. J. (1999). Authorship: "Law" and order. *Academic Emergency Medicine*, 6(4), 297–301.

Grafton, A. (1997). *The Footnote: A Curious History*. Cambridge, MA: Harvard University Press.

Gross, A. G., Harmon, J. E., and Reidy, M. S. (2000). Argument and 17th-century science: A rhetorical analysis with sociological implications. *Social Studies of Science*, 30(3), 371–396.

Harsanyi, M. A. (1993). Multiple authors, multiple problems: Bibliometrics and the study of scholarly collaboration. A literature review. *Library & Information Science Research*, 15(4), 325–354.

Heffner, A. G. (1979). Authorship recognition of subordinates in collaborative research. *Social Studies of Science*, 9, 377–384.

Heffner, A. G. (1981). Funded research, multiple authorship, and subauthorship collaboration in four disciplines. *Scientometrics*, 3, 5–12.

Hoen, W. P., Walvoort, H. C., and Overbeke, A. J. P. M. (1998). What are the factors determining authorship and the order of the authors' names? A study among authors of the *Nederlands Tijdschrift voor Geneeskunde* (*Dutch Journal of Medicine*). *Journal of the American Medical Association*, 280, 217–218.

Horton, R. (1998). The unmasked carnival of science. *The Lancet*, 351(9104), 688–689.

Houston, P. and Moher, D. (1996). Redundancy, disaggregation, and the integrity of medical research. *The Lancet*, 347(9007), 1024–1026.

Hyland, K. (1999). Academic attribution: Citation and the construction of disciplinary knowledge. *Applied Linguistics*, 20(3), 341–367.

International Committee of Medical Journal Editors. (1997). Uniform requirements for manuscripts submitted to biomedical journals. *Journal of the American Medical Association*, 277, 927–934.

Johns, A. (2003). The ambivalence of authorship in early modern natural philosophy. In: Biagioli, M. and Galison, P. (Eds.). (2003). *Scientific Authorship: Credit and Intellectual Property in Science*. New York: Routledge, 67–90.

Kahn, K. S., Nwosu C. R., Khan, S. F., Dwarakanath, L. S, and Chien, P. F. (1999). A controlled analysis of authorship trends over two decades. *American Journal of Obstetrics and Gynecology*, 181(2), 503–507.

Kassirer, J. P. and Angell, M. (1991). On authorship and acknowledgements. *New England Journal of Medicine*, 325(21), 1510–1512.

Katz, J. S. and Martin, B. R. (1997). What is research collaboration? *Research Policy*, 26(1), 1–18.

Katzen, M. F. (1980). The changing appearance of research journals in science and technology: An analysis and a case study. In: Meadows, A. J. (Ed.). *Development of Scientific Publication in Europe*. Amsterdam: Elsevier, 177–214.

Kernan, A. (1990). *The Death of Literature*. New Haven, CT: Yale University Press.

Kevles, D. J. (1998). *The Baltimore Case: A Trial of Politics, Science, and Character*. New York: Norton.

King, J. T. (2000). How many neurosurgeons does it take to write a research article? Authorship proliferation in neurosurgical research. *Neurosurgery*, 47(2), 435–440.

Kitcher, P. (1993). *The Advancement of Science: Science Without Legend, Objectivity Without Illusions*. New York: Oxford University Press.

Klein, C. J. and Moser-Veillon, P. B. (1999). Authorship: Can you claim a byline? *Journal of the American Dietetic Association*, 99, 77–79.

Kling, R. and McKim, G. (2002). Not just a matter of time: Field differences in the shaping of electronic media in supporting scientific communication. *Journal of the American Society for Information Science*, 51(14), 1306–1320.

Kling, R., McKim, G., Fortuna, J., and King, A. (2000). A bit more to IT: Scientific multiple media communication forums as socio-technical interaction networks. Available at www.slis.indiana.edu/SCIT.

Knight, D. (1976). *The Nature of Science: The History of Science in Western Culture Since 1600*. London: Deutsch.

Koehler, W. et al. (1999). A bibliometric exploration of the demographics of journal articles: Fifty years of *American Documentation* and the *Journal of the American Society for Information Science*. Available at www.ou.edu/cas/slis/courses/Methods/jbib/.

Kunst, F. et al. (1997). The complete genome sequence of the Gram-positive bacterium Bacillus Subtilis. *Nature*, 390(6657), 249–256.

Laband, D. N. and Tollison, R. D. (2000). Intellectual collaboration. *Journal of Political Economy*, 108(3), 632–662.

Latour, B. and Woolgar, S. (1979). *Laboratory Life: The Social Construction of Scientific Facts*. Beverly Hills, CA: Sage.

Manguel, A. (1997). *A History of Reading*. London: Flamingo.

Manten, A. A. (1980). Development of European scientific journal publishing before 1850. In: Meadows, A. J. (Ed.), *Development of Scientific Publication in Europe*. Amsterdam: Elsevier, 1–22.

McCain, K. W. (1991). Communication, competition, and secrecy: The production and dissemination of research-related information in genetics. *Science, Technology, & Human Values*, 16(4), 491–516.

McDonald, K. A. (1995, April 28). Too many co-authors? *Chronicle of Higher Education*, A35–36.

Merton, R. K. (1973). The normative structure of science. In: Merton, R. K., *The Sociology of Science: Theoretical and Empirical Investigations*, edited and introduced by N. W. Storer. Chicago: University of Chicago Press, 267–278.

Montgomery, L. (2003). *The Chicago Guide to Communicating Science*. Chicago: Chicago University Press.

Mullins, N. C. (1973). *Theories and Theory Groups in American Sociology*. New York: Harper & Row.

Newman, D. (1996). Writing together separately: Critical discourse and the problems of cross-ethnic co-authorship. *Area*, 28(1), 1–12.

Nowak, M. A. and Sigmund, K. (2000). Shrewd investments. *Science*, 288, May 5, 819–820.

Patel, N. (1973). Collaboration in the professional growth of American sociology. *Social Science Information*, 12(6), 77–92.

Price, D. J. de Solla. (1963). *Little Science, Big Science*. New York: Columbia University Press.

Rayward, W. B. (1992). Restructuring and mobilizing information in documents: a historical perspective. In: Vakkari, P. and Cronin, B. (Eds.). *Conceptions of Library and Information Science: Historical, Empirical and Theoretical Perspectives*. London: Taylor Graham, 50–68.

Relman, A. S. (1983). Lessons from the Darsee affair. *New England Journal of Medicine*, 308, 1415–1417.

Rennie, D. and Flanagin, A. (1994). Authorship! Authorship! Guests, ghosts grafters, and the two-sided coin. *Journal of the American Medical Association,* 271(6), 469–471.

Rennie, D., Yank, V., and Emanuel, L. (1997). When authorship fails: A proposal to make contributors accountable. *Journal of the American Medical Association*, 287(7), 579–585.

Rennie, D. and Yank, V. (1998). If authors became contributors, everyone would gain, especially the reader. *American Journal of Public Health*, 88(5), 828–830.

Riesenberg, D. and Lundberg, G. D. (1990). The order of authorship: Who's on first? *Journal of the American Medical Association*, 264, 1857.

Saffran, M. (1989). On multiple authorship: Describe the contribution. *Scientist*, 3(6), 9.

Shapin, S. (1995). *A Social History of Truth: Civility and Science in Seventeenth-Century England*. Chicago: University of Chicago Press.

Slone, R. M. (1996). Coauthors' contributions to major papers published in the *AJR*: frequency of undeserved authorship. *American Journal of Roentgenology*, 167, 571–579.

Smith, R. (1997). Authorship is dying: Long live contributorship: The *BMJ* will publish lists of contributors and guarantors to original articles. *British Medical Journal*, 315(7110), 696.

Spiegel, D. and Keith-Spiegel, P. (1970). Assignment of publication credits: Ethics and practices of psychologists. *American Psychologist*, 25, 738–747.

Squires, B. P. (1996). Authors: Who contributes what? *Canadian Medical Association Journal*, 155(7), 897 898.

Steiner, G. (2003). *Lessons of the Masters*. Cambridge, MA: Harvard University Press.

Stern, E. B. (2000). Authorship criteria: Opening a dialogue. *American Journal of Occupational Therapy*, 54(2), 214–217.

Traweek, S. (1992). *Beamtimes and Lifetimes: The World of High Energy Physics*. Cambridge, MA: Harvard University Press.

Verner, D. A. (1993). Astronomy acknowledgement index 1992. *Messenger*, 71, 59.

Wilcox, L. J. (1998). Authorship: The coin of the realm, the source of complaints. *Journal of the American Medical Association*, 280(3), 216–217.

Chapter Four

Information Space

In their landmark book, *Laboratory Life*, Latour and Woolgar (1979, 88) describe how "rats had been bled and beheaded, frogs had been flayed" in the service of science. They portray laboratory activity at the Salk Institute "as the organization of persuasion through literary inscription." The end result of the spilling of blood and guts is the journal article, a sanitized and depersonalized artifact far removed from the messiness of what went before: "A laboratory," they aver, "is constantly performing operations on statements; adding modalities, citing, enhancing, diminishing, borrowing, and proposing new combinations" (Latour and Woolgar 1979, 86–87). Ultimately, these literary operations mask the literal operations carried out on experimental rats by scientists in wet labs. We shouldn't, of course, be surprised by any of this; "In scientific articles, reason proceeds down a high road that leads from darkness to light with not the slightest error, nor a hint of a bad decision, no confusion, nothing but perfect reasoning" (Rheinberger 2003, 315). Finely crafted articles also draw a convenient veil over "research scribbling," a set of integral activities ranging from laboratory sketching to trying out calculations that lies in the space between "the materialities of the experimental systems and the various written communications that are eventually released to the scientific community" (Rheinberger 2003, 314).

It occurs to me, and I say this without positing any form of strategic silencing of the social on the part of scientometricians, that the messiness of professional and human relations is elided in similar fashion from most standard bibliometric and scientometric accounts of knowledge growth, diffusion, and sharing. "Never mind the underlying social reality, feel the data," seems to be the unspoken credo. Sociological relativists, by the same token, are characterized as being antipathetic to grand narratives and large-scale number crunching. Take the case of citation, an integral part of scientific writing, yet one that is sometimes dismissed as necessarily subjective and arbitrary (see Cronin [1984] for a review of the arguments), and thus an inherently unreliable basis for mapping the structure of disciplines or for evaluating scientific performance. But, although I want to steer a careful course between the Scylla of uncritical normativism and the Charybdis of naive constructivism, I find it hard not to go along with White's (1990, 9) pragmatism in the matter of citations and what they tell us about scientific communication: "[w]hen one sees that scores, hundreds, and even thousands of citations have accrued to a work, an author, a set of coauthors, it is difficult to believe that all of them are suspect. Why not believe there is a norm in citing . . . and that the great majority of citations conform to it?"

Support for the so-called normative position comes from two noteworthy studies of recent origin. Kurtz and colleagues at the Harvard-Smithsonian Center for Astrophysics (Kurtz et al. 2004) have used the NASA (National Aeronautics and Space Administration) Astrophysics Data System to compare the obsolescence function as measured by "reads" of records in the system with the obsolescence function as measured by citations. Their statistical analyses show that reads and citations "fundamentally measure the same thing, the usefulness of an article." The authors conclude that this "proves that the normative theory of citation is true in the mean." Pertinently, Kurtz et al. go on to say that "the private act of reading an article entails none of the various sociological influences that the public act of citing an article does [which] suggests that in the mean these factors do not influence the citation rate." Citation, in other words, is about what one knows, not whom one knows. These results confirm the earlier findings of Baldi's (1998, 843) multivariate analysis of citations in the literature of celestial masers, a subfield (coincidentally) of astrophysics.

Authors are most likely to cite articles that are relevant to their work in terms of subject, recency of knowledge, theoretical orientation, and seem to have little concern with the characteristics of the authors who write them. This finding suggests that, at least in the research area examined, one's position in the stratification structure of science is likely to be the result of the worth and usefulness of one's scientific contributions rather than the reverse, as social constructivists would have us believe.

One of the many measures used by Baldi was the existence of social ties, which he operationalized as potentially citing or potentially cited articles' authors having "ever worked at the same institution or received their Ph.D. from the same graduate department" (1998, 837). To his credit, Baldi does not think that this is the end of the story. First, caution is needed in generalizing from this single-domain study to other scientific disciplines, not to mention humanistic fields where there may be considerably less consensus over what constitutes high-quality work. Second, he concedes (1998, 844) that more work is needed "to elaborate the *kinds* of social ties that matter" and that "[f]uture studies should assess the effects of various social relationships among authors on citation occurrence by collecting more detailed information on type of ties."

It is important to state that a commitment to the normative view of citation does not necessarily preclude or somehow invalidate one's having an interest in the social and biographical minutiae of scholarly communication. For example, Mählck and Persson (2000, 81, 84) introduced the term "socio-bibliometric mapping" and showed that "if you collaborate with a certain author, you will tend to cite him or her." Rowlands (1999, 543) has also produced evidence of "convergence between recognition, social ties and collaborative forms of activity." But I want to take this one step further. A simple, self-referential illustration may help. In the course of writing this chapter, I shall cite, amongst many others, Eugene Garfield and Elisabeth Davenport, both of whom I have known personally for two decades. The former is best described as a professional friend (for example, I co-edited a *Festschrift* in his honor), the latter as a frequent collaborator, coauthor, and close personal friend. I don't cite them *because* of our social ties, but because their ideas are relevant to my work. At the same time, the odds on their being cited by me are increased as a result of the preexisting social ties; I know them and their publications well; I interact with them, exchanging thoughts and materials; I have ready access to them and their resources, broadly defined. In the case of the latter, we have been active, and occasionally collocated, collaborators for many years. A consequence of my citing Garfield and Davenport is that it reduces the likelihood of others in the citable author pool from being selected. All other things (the citable work's topicality, relevance, currency, etc.) being equal, strong social ties will presumably trump weak or nonexistent ties (see Granovetter [1973, 1983] for a general discussion of weak and strong ties and Burt [1992] on "structural holes," the connectivity gaps between discrete social or professional groups). Call it preferential attachment, a statistical fact of not only scholarly but also social life.

Physical proximity, I would have to say, has also played a large part in my including citations to the information visualization research of my local colleague, Katy Börner. Were our offices not but ten meters apart, I'm not sure that she would have featured here; contiguity helped spark informal exchanges and a partial convergence of research interests. But, equally, I cite others (the aforementioned Baldi and Kurtz being cases in point) whom I have never met and with whom I have no social or personal connections whatsoever. Here, they are cited purely because their work is materially and in some ways uniquely relevant to my argument. I could go through the entire list of references at the chapter's end classifying the nature of the relationship I have with each cited author but, for now, I trust you follow my drift. The significance of the social can be seen even more clearly from another angle. Four of the individuals mentioned in the acknowledgments at the front of this book helped shape the ideas in this chapter. All four are known to me personally; one is a close friend and immediate colleague, two others are colleagues and friends, and the fourth is, as already mentioned, a long-standing personal friend and frequent coauthor. On this admittedly limited and parochial basis, it would be churlish to deny that social and biographical ties play a part in the shaping of one's intellectual choices and trajectory, which is why I am comfortable describing academic writing as an instance of distributed cognition (see chapter 5, "Intellectual Collaboration"). Allowing that is not, however, tantamount to privileging the constructivist (interpretivist) position over the normative (structuralist) position, a point well made by Borgman and Furner (2002, 46).

> At a general level, the interpretivist and structuralist trends are oppositional, in that the former is characterized by an emphasis on the primacy of the citer's personal actions (influenced but not determined by context), whereas the latter consists of the renewal of interest in identifying probabilistic regularities and patterns that (as some might be tempted to say) "govern" human behavior. This should not be taken to imply, however,

that a reconciliation is logically impossible, or even that evidence of both trends cannot be found simultaneously in single studies.

In my opinion, a little social archaeology would not go amiss within the expanding literature of scientometrics, informetrics, and cognate fields but, as we shall see, the data-gathering processes associated with this kind of activity are often labor intensive; there is as yet no direct equivalent of the *Science Citation Index* recording the parallel webs of personal, social, and affective ties that bind communities of scholars and researchers to which we can turn for large-scale data processing. The work involved in uncovering the unseen is painstaking, not unlike an archaeological dig. For the present, we rely on shards and shadows to tell the story of informal scholarly collaboration

WARM BODIES AND COLD FACTS

A decade ago, I coauthored a bagatelle in the *Journal of Information Science*, tongue-in-cheekily entitled "What's the Use of Citation?" (Snyder, Cronin, and Davenport 1995). Here I am trying to adopt a somewhat similar posture by stepping back a little and looking at my primary intellectual community—information science and scientometrics, I'll label it for convenience—with the detachment of an outsider. Specifically, I want to consider the institutional arrangements and social ties that are often imperceptibly etched in the frequency distributions and citation maps that emerge from our research endeavors. In doing so, I visit research literatures that at first glance might seem beyond the scientometric pale. I want to know whether invisible colleges, virtual (academic) communities, coauthorship patterns, cocitation maps, and author networks are first and foremost semantic representations or, rather, statistical agglomerations of "authors known to each other as warm bodies rather than as labels on literature," as Price (1970, 4) memorably phrased it. Of course, one intuitively "knows" that these various representations reflect an admixture of social (professional and personal) and semantic (intellectual and scientific) influences—socio-scientometrics would be a cumbersome if appropriate coinage—but one has the sense that the data are supposed to speak for themselves and that the social dimensions of scientometrics are neither relevant nor, perhaps, of too much intrinsic interest to most members of the International Society for Scientometrics and Informetrics (see www.issi-society.info/).

Half a century ago, Robert Merton (sociology) and Joshua Lederberg (genetics) were quick to grasp the wider sociological and science policy significance of Eugene Garfield's developmental work in citation indexing (Cronin and Atkins 2000). As Wouters (2000, 69) has noted, the advent of computerized citation indexes "created a completely novel symbolic universe, based on the semiosis of the citation. Garfield turned lead (the reference) into gold (the citation) in his search for the philosopher's stone." By way of an aside, I'm surprised Wouters didn't come up with the coinage "*Sign*tometrics" to mobilize his observation that "scientometrics is the practice of creating a new representation of science . . . using a new system of signs on science" (Wouters 1993, 12). More on this subject and the relationship between "reference" and "citation" can be found in chapter 7 ("Symbolic Capitalism").

As I write, it's almost fifty years to the day since the gold rush began. Bibliometric pioneers such as Derek de Solla Price (physics and history of science) and Belver Griffith (psychology) were unusually talented interpreters of the data they marshaled. Henry Small (history of science) and Howard White (library and information science) follow firmly in that tradition, but they are in a distinct minority. My quartet of then-and-now grandees (Price, Griffith, Small, and White) was possessed of a sociological imagination, in addition to being numerate; moreover, they all write (or wrote) with panache. They managed to vivify their potentially arid material to great effect. But for every Price or Small, there is a legion of lesser names. These individuals may stand on the shoulders of giants but, ironically, they seem not to be able to see more clearly as a result. Even the field's eponymous journal, *Scientometrics*, for all the excellent efforts of its distinguished editor, Tibor Braun, still occasionally provides a resting place for papers that ought not to have seen the light of day. The same criticism can, of course, be leveled against the many other journals (e.g., *Journal of the American Society for Information Science and Technology, Journal of Information Science*) that accept biblio-, infor-, and scientometrically inclined papers (and, yes, I am guilty of authoring some of those supernumerary publications). Counting for counting's sake seems to be the raison d'être of at least some would-be informetricians, but solid progress is nonetheless being made on a number of fronts (Wouters and Leydesdorff 1994).

As a disciplinary tribe, we sometimes fail to see the wood for the trees—unlike our lustrous predecessors, who saw the wood before most of the trees had even been planted. The aforementioned grandees obsessed less about the finer points of mathematical analysis or statistical modeling (important though these activities indubitably are and adept though our founding fathers were at presenting and interpreting bibliometric data; Price, we are told [Merton and Garfield 1986, 2], was blessed with "ample numerical imagination") than about developing scientometrics as a robust meta or second-order field, one that contributes to social studies of science, and, indeed, the historiography of science. Price, as Furner (2003, 13) tells us, was interested in "the study of social factors in science in general" and in advancing both scientometric and sociological theory. Intellectual expansiveness and ambitiousness are clearly visible in Price's justifiably much-cited writings, but these virtues are generally less in evidence today, despite the quite striking growth of interest in both classical bibliometrics and its cyber derivatives (see Thelwall, Vaughan, and Björneborn 2005). A plausibly charitable explanation for the lack of integration and synthesis would be that the field has grown and continues to expand at such a rate that a unified account constitutes an unreasonable expectation.

INFORMATION SPACES

I now revisit and briefly review some early and also more recent research in information science to demonstrate its potential relevance to current thinking on spatiality and social interaction. In the process, I draw upon a range of literatures in order to show topical connections that are sometimes overlooked. I begin with physical space (how distance influences the degree of informal communication between scientists/scholars), move on to intellectual space (conceptual/cognitive mapping of science), and conclude by suggesting, optimistically it must be said, that hybrid spatial analyses (combining social and intellectual dimensions) are needed to faithfully capture the multiplex nature of scholarly communication. I raise a number of related questions: (a) How are social relations—forged largely in physical or "real-space" (Dourish 2001a, 2001b)—reflected in the architecture of intellectual, or semantic, space? (b) Do cocitation maps of science depict purely intellectual networks or webs of sociocognitive interactions? How, to put it otherwise and to paraphrase Suchman (1987), might we represent (socially) situated semantics? (c) Are there paratextual indicators, apart from citation and coauthorship data, that could be used to map sociocognitive ties? and (d) Can social ties be practicably captured and visualized?

Notions of information space abound. For example, sociologists, urban planners, and economists are interested in the geography of the information economy, the emergence of the postindustrial city and the ways in which developments in internetworking have created a new spatial logic, one based on interactions between the space of information flows and the space of places (Castells 2000; Hepworth 1989; Zook 2003). In a more abstract vein, Boisot's (1995, 1999) integrated conceptual framework, the I-Space (Information-Space), creatively explores the relationships between the processes of information codification, abstraction and diffusion in social spaces. The I-Space (instantiated in his ubiquitous, morphing cube) allows him to show how different organizational structures and institutional cultures condition the nature and efficacy of information exchange—and perceptions of information value—and how those structures, in turn, exert a reciprocal influence on information flows: a process of structuration, in other words (Boisot 1995, 6).

Within information retrieval (an important subfield of information science), the vector space model—most closely associated with the work of Salton—is a well-established means of representing the similarity between documents based on weighted term frequency counts; the more topically similar two documents are, the closer they will be located to one another in concept/semantic space (Salton, Wong, and Yang 1975).

Information science, notably the subfield of bibliometrics, has long been interested in representing and mapping intellectual space. Recent advances in information visualization techniques and related statistical methods have greatly facilitated the depiction of semantic spaces and networks of scholarly interaction (Mackinlay, Rao, and Card 1995). The lexicon of information visualization is spatially explicit and metaphor rich—maps, landscapes, terrain, peaks, and cities—and the use of cartographic techniques to show the relationships among groups of scholarly publications is well established (Börner, Chen, and Boyack 2003; Skupin 2002).

According to Heimeriks and van den Besselaar (2002, 11), bibliometric maps of science "are landscapes of scientific research fields created by quantitative analysis of bibliographic data. In such maps the 'cities' are, for instance, research topics. Topics with a strong cognitive relation are in each other's vicinity and topics with a weak relation are distant from each other." Skupin, a geographer by trade, has acknowledged the early contributions of information science—Paul Otlet in particular—to the use of cartography in the information visualization domain.

> Map metaphors have been associated with the handling of non-geographic information for a long time. They can be traced as far back as the late 19th century when Paul Otlet, regarded by many as the father of information science, made explicit reference to mapping of intellectual domains. Otlet envisioned the use of maps in the exploration of unknown information terrain and even pondered the role of scale in such exploration. (Skupin 2000, 91)

Mapping techniques are now routinely used in information science to reveal the intellectual structure of disciplines (White and McCain 1998) and to create real-time, interactive author maps (Lin, White, and Buzydlowski 2003). They have also been used for social network analysis (Koku, Nazer, and Wellman 2001; Matthiessen, Schwartz, and Find 2002). Most recently, White (information science) and Wellman (social network analysis) have collaborated to explore the sociocognitive structure of specialty groups (White, Wellman, and Nazer 2004). However, there is important prior work to be found in the literature of bibliometrics and scientometrics, starting with Price's (1965) landmark article, "Networks of Scientific Papers," one of the earliest attempts to use citation data to reveal the intellectual structure of a scientific subfield. As it happens, small-world theorists have lately come to recognize the importance of the citation network: "all scientific publications are part of a web of science in which nodes are research publications connected by citations" (Barabási 2002, 169). The post–World War II Anglo-American information science community clearly grasped the nature and significance of informational spaces, both literal and figurative. However, this fact has sometimes been overlooked.

Thomas Allen carried out groundbreaking studies on the problem-solving and communication behaviors of engineers and scientists and the role of the technological gatekeeper in facilitating information flow. Writing more than three decades ago in the *Annual Review of Information Science and Technology*, he noted that "[c]ommunication probability decreases with the square of distance and . . . reaches its asymptotic level within 25 yards" (Allen 1969, 11). Other kinds of studies have found that there is a strong relationship between distance and the amount of use made of information sources and facilities. The resources (human or documentary, physical or virtual) we routinely use in our work lives are not necessarily the best or most potentially useful for the task in hand; rather (and to oversimplify), we engage in a continual trade-off between factors such as convenience, accessibility, cost, and quality (King et al. 1984). This, of course, has implications for architects, interaction designers, database producers, and others. But there are systemic—not just task-level—ramifications. For example, Odlyzko (2002) maintains that the shift from print-based to digital modes of scholarly communication is ineluctable. He offers the following reasons: (1) the present system is inefficient, expensive, and inflexible; (2) scholars are engaged in a "war for the eyeballs," a fact of life in the so-called attention economy; (3) a critical mass of material is available on the web; and (4) a web presence translates into more eyeballs. Thus, self-interest dictates that scholars migrate to the web. Odlyzko's other pertinent observation is that readers will settle for "near substitutes" or "inferior forms of papers" if these can be reached easily.

As was noted in chapter 2, Odlyzko invokes Christensen's (2000) notion of disruptive technologies to explain the demise of print but, equally, he could have cited Gladwell's (2002) idea of the "tipping point," which has the added merit of foregrounding the role played by opinion shapers (Christensen's framework operates at the level of structural dynamics, not individual influencers). Gladwell identifies and characterizes three kinds of key social actors—mavens, connectors, and salesmen (2002, 70): "Mavens are data banks. They provide the message. Connectors are the social glue: they spread it." His third category, salesmen, is "a select group of people . . . with the skills to persuade us when we are unconvinced of what we are hearing." These social pollinators perform a variety of functions, some of which are similar to those carried out by the technological gatekeepers and boundary spanners described by Allen and others in the 1960s and 1970s.

PHYSICAL PRESENCE AND PROXIMITY

Contrary to popular belief, the emergence of virtual networks and on-line communities does not negate the importance of physical proximity in the conduct of research, just as online trading does not mean the dematerialization of commerce or the death of real estate. Presence matters in both real and virtual space. Confirmatory evidence comes in a variety of forms and from a variety of contexts. Katz (1994, 31) found that the number of intranational collaborations "decreases rapidly as a function of distance separating research partners," while Thelwall has shown that the "extent of academic web site interlinking between pairs of U.K. universities decreased with geographic distance . . . neighboring institutions were very much more likely to interlink than average" (Thelwall, Vaughan, and Björneborn 2005, 106). Knorr Cetina (1999, 212) notes that "collaborators not continually at CERN . . . feel left behind, as if they were chasing after something that is always two steps ahead. In principle, almost everything is accessible to everyone at all times—but in practice, information circulates through local discourse at the center, which one must be physically plugged into . . . to be up to date." To be where the action is, if I may appropriate the title of Dourish's (2001a) book, counts.

Humans have a penchant for minimizing resistance and economizing on effort. George Kingsley Zipf's "Principle of Least Effort" and Allen's "30-Metre Principle" neatly encapsulate such near-universal "satisficing" behaviors. By way of an aside, Allen (1969) went on to note that the architectural profession seemed to have ignored the spatial determinants of social interaction (how distance affects frequency of informal communication). Physical connectivity and degrees of presence matter, even in an age of electronic publication, online fora, and digital networks. Geographers recognize that the space of flows and the space of places are co-constitutive; the geography of the Internet is a case of intertwined worlds, the physical and the virtual (Zook 2003). In similar vein, media complementarity (not substitution) is a defining feature of contemporary scholarly communication and, indeed, life more generally (see Haythornthwaite and Harger [2005] on the meshing of offline geography and online interactions). E-mail has not displaced other means of information exchange and interaction; in fact, e-mail use correlates positively with most traditional forms of communication (face-to-face, phone, fax, and mail).

Although the Internet helps scholars to maintain ties over great distances, physical proximity still matters. Those scholars who see each other often or work nearer to each other email each other more often. Frequent contact on the internet is a complement for frequent face-to-face contact, not a substitute for it. (Koku, Nazer, and Wellman 2001, 1750)

Social relations are subtly inscribed in the architecture of the scholarly communication system. Many academic domains/specialty groups combine near year-round conference caravanning with intensive electronic communication via e-mail, listserv, chat rooms, and so forth. Virtuality has not diminished the importance of in-group membership. But changes have occurred. The invisible colleges described more than three decades ago by Crane (1972) are being progressively reconstituted as "visible colleges" (Koku, Nazer, and Wellman 2001). They have been "outed." Today, digital communications media make it much easier to track sociometric stars than in Crane's or Price's day; we can see who posts, blogs, and responds; who generates a buzz and attracts attention online. Current efforts to deal with the problems of "teleidentity" and visualize social environments and patterns of interaction, such as those conducted by the Sociable Media Group at MIT (Boyd, Lee, Ramage, and Donath 2002), may help increase communicative transparency in certain instances, but the traces left in the ether don't describe the full spectrum of social relations that structure specialty groups, even if they reveal some of the institutional and professional circuitry.

CITATION LINKS AND HYPERLINKS

Garfield's (1955) idea of indexing the literature of science—described in his seminal *Science* paper—by the material that is cited by that literature was elegantly simple. Early developments in automated information retrieval systems made increasingly ambitious bibliometric analyses possible (Cronin and Atkins 2000). Garfield was interested in the visual representation of citation patterns (the unseen lattices of references implicit in the citation databases; for a bibliography, see www.garfield.library.upenn.edu/mapping/mapping.html). The pioneering work of his ISI colleague Henry Small stimulated widespread interest in using aggregated citation data to chart the emergence and evolution of scientific specialties (Small 1999). Such visualizations (for instance, the early *Atlas of Science* covering biochemistry and molecular biology) provide users/readers with "a mental image of a domain space" (Williams, Sochats, and Morse 1995, 163), without which it would be difficult to make sense of the data or infer the overall structure of a domain or specialty group.

Hundreds of disciplinary snapshots and maps (using authors and documents) of intellectual spaces have been generated as a result of the foundational work of Price, Garfield, and Small yet, as White and Mc-Cain (1997, 116) note, "the principal bibliometricians have never been good, journalistically speaking, at placing what they do in a new, widely intelligible framework." However, ISI's contributions to scientific mapping of knowledge spaces, its "graphical legacy" (Davenport and Cronin 2000, 528), has not gone unrecognized; the developers of the Google search engine—Sergey Brin and Lawrence Page—have publicly acknowledged the similarities between hyperlinking and citation linking. The parallels between citations and hyperlinks have also been explored by Borgman and Furner (2002, 13–14).

> The success of systems such as Google and Clever is reflected in the extent to which research in Web-based IT is currently dominated by attempts to implement link-analytic techniques at ever-increasing levels of sophistication. It should be recognized (in a manner that Brin and Page, for example, do not) that bibliometricians working on conventional citation analysis have, in their efforts to replace reliance on crude citation counts, produced similar formulations at earlier dates.

Even though the web is very different from the controlled document collections that underpin ISI's citation indexes (Page et al. 1998), there are functional equivalents, if one thinks of links (hyper and citation) as being akin to votes, judgments, pointers, recommendations, expressions of trust, or endorsements. For an early discussion of the potential benefits of hyperlinked citations, see Davenport and Cronin (1990, 177). As it happens, those masters of physical space—architects—have not been impervious to the value of mapping intellectual space. Mitchell noted presciently in his foreword to Bonto's (1996, x) indexical/citation analysis of American architects and texts that "interest in a subject can be measured by the number of hyperlinks pointing to WWW pages."

The rationale underpinning citation mapping has been expressed succinctly by Small (1973, 265): "If it can be assumed that frequently cited papers represent the key concepts, methods, or experiments in a field, then co-citation patterns can be used to map out in great detail the relationships between these key concepts." The reality of science emerges—is constructed—from these representations (Wouters 1999, 129). This, of course, invites the following questions: are the resultant maps reflective of purely ideational interactions; are citation behaviors normatively governed or idiosyncratic in nature (see Cronin [1984] for a review of the relevant literature on citer motivation)? More specifically, to what extent do social and psychological factors influence the selection of citations by an author—the citations that become the coordinates of intellectual space? To what extent can social and affective relations be imbricated on such maps? To what extent should the semantic spaces described by citation maps be viewed as *socio*cognitive spaces?

CITATION IDENTITY AND CITATION IMAGE

White (2001) has undertaken a study of scholars' *recitation* patterns, analyzing the identities of those authors who are cited recurrently by a given author over time. Working with a small sample of information scientists, he attempted to tease out some of the social, collegiate, and institutional ties that might influence citation (i.e., intellectual) choices. This kind of authorial exegesis is only possible if one is familiar with both the subject domain and the focal set of authors. White concludes (2001, 93) that most members of his sample are "affected by social networks—that is, they cite authors whom they know personally from school, the workplace, or an invisible college (defined as researchers with similar interests who communicate and collaborate although their institutional bases differ and are possibly far apart)." Thus, one's location in intellectual space to some extent reflects one's place in the physical world.

We (Cronin and Shaw 2001) subsequently employed a variant on White's approach, looking closely at the citation identities (based on twenty years' worth of ISI data) of three information scientists, A, B, C, whom we knew extremely well. We constructed their citation identities (those whom they cited) and their citation images (those who cited them) and looked at the top twenty-five names on each list. To give a flavor of the findings, *A*'s second most highly cited author was a former doctoral student and faculty colleague. *A* was most frequently cited by this protégé. *A* knew personally fourteen of the twenty-five names, and had shared an office for two years with one. Among the most frequently cited names by *B* were four or five former colleagues from another institution; *B*, in turn, was most highly cited by a former colleague. *B*'s identity included an erstwhile doctoral student. *B*, to the best of our knowledge, personally knew almost all of those whom he cited and, in turn, likely knew most of those who cited him. *A* featured midway down *B*'s image-maker list. *C* most frequently re-cited his doctoral supervisor, who cited him in turn. *C*'s identity list included a former doctoral student. *A* also featured on *C*'s image-maker list. Late in their careers, *A*, *B*, and *C* converged on the same academic department, and the level of mutual referencing increased considerably thereafter. One could go on. The mesh of collegiate and mentor-advisee relationships may be imperceptible to most observers, but it is nonetheless real. Physical place and sociality clearly play a role in shaping intellectual spaces, but how might we go about visualizing these social linkages, both weak and strong?

Both the aforementioned studies suggest that physical presence and professional familiarity are factors that influence citation behavior, though, of course, personal acquaintance is neither a necessary nor sufficient reason to cite an author. That said, the work of one's associates and friends might be more accessible and no less pertinent than the work of others, and thus warrant citation on merit alone. Naturally, we would expect social ties to influence citation behavior to some extent. One of the reasons for working together (for setting up collaborations, local or distributed) is to build a common cause, combine talents and resources, energize a collective research agenda, or mobilize a particular paradigm or *Weltanschauung*; in other words, to achieve some kind of multiplier effect, or what economists refer to as returns to specialization. That, of course, is not to say that colocation is synonymous with groupthink. In the illustration above, *B*, for instance, was part of a widely recognized intellectual group, or "school," but he also had a strong intellectual trajectory of his own. Indeed, *B* was a boundary spanner/gatekeeper par excellence (more so than either *A* and *C*), as we found when we identified the many fields and disciplines from which he harvested citations.

Being together in a shared space, or meeting face-to-face, shapes the intellectual spaces in which scholars' work is located. The central issue has been formulated thus: "Is it primarily *who* citers know (social structure) or *what* they know (intellectual structure)?" (White, Wellman, and Nazer 2004, 111). Of course, things are not quite that simple, as White et al. (2004, 125) concede: "Who you know pays off only if the people you know have something worth knowing—something plainly relevant to your own claims." But unraveling the social dimensions of citation links is still not the whole story. Citations are the most visible part of the influence iceberg; acknowledgments are the part largely hidden below the water line. Scholastic debts are routinely recorded in a variety of ways; citations just happen to be relatively easy to capture and display. In short, citation-based maps are necessarily partial representations of scholars' interactions.

The support networks implicitly described in acknowledgments reinforce the idea that research and scholarship are socially embedded rather than purely individualistic activities. Acknowledgments bear witness to the myriad ties—social, technical, intellectual, affective—that bind scholars together. Indeed, academic writing, as I argue in the next chapter, is a compelling instance of distributed cognition (Hutchins 1995), or, to use Clark's (2003, 8) phrase, of "an extended cognitive system." The growth in acknowledgments was an important, if underappreciated, development in scholarly publication during the second half of the twentieth century (Cronin 1995; Cronin, Shaw, and La Barre 2003, 2004). In fields as diverse as astrophysics, chemistry, cell biology, and psychology, almost every scientific paper includes (often substantive) expressions of gratitude to a distributed population of peers, informal collaborators, and trusted assessors; individuals whose material contributions and/or conceptual inputs have apparently made a difference to the work being reported. Scientists are connected, sociotechnically, and with the progressive "collectivization of academic science" (Etzkowitz and Kemelgor 1998)—what Ziman (2000) has termed "post-academic" science—those distributed sociotechnical networks will both increase and intensify. Or, as Collins (2003, 150), echoing Clark, put it: "Rapid-discovery scientists have always owed their reputations, and their ability to make new discoveries, to being connected to evolving generations of equipment. . . . Modern science has always been a cyborg network in this sense."

Collaboration in science is often investigated via coauthorship, yet coauthorship data do not comprehensively capture scholars' functional interdependence or the structural complexity of "the socio-cognitive network" (Merton 2000, 437). If the intellectual structure of fields is to be mapped with fidelity, then acknowledgment data should, strictly speaking, be taken into account. More specifically, we would want to look at the geographic spread of trusted assessors—Davenport (1993, 310) identifies levels of community: street, parish, national—to further determine the influence of physical space on patterns of both intra- and interdisciplinary collaboration.

CONCLUSIONS

The conceptual spaces defined by authors' citation behaviors reflect the importance (modest or otherwise) of physical proximity and personal acquaintanceship in shaping patterns of information exchange and knowledge diffusion. It is clear that physical proximity does play some role, direct or indirect, in molding citation behavior, even if the conduits and connections are typically unseen to all but privileged insiders or bibliometric researchers. Of course, it is hardly surprising, to return to our earlier example, that if *A*, *B*, and *C* are collocated in the same department or laboratory, and have cognate or converging research interests, then those latent ties will become manifest, resulting in greater intragroup exchange of resources, ideas, and citations over time. A shared physical site (lab, office complex) or virtual space (collaboratory, listserv) can act as an incubator or accelerator of ideas, which, of course, is precisely why we attend conferences and professional meetings. One is not here referring to the forging of purely personal, or "content-neutral" (White et al. 2004, 125) ties, but the natural codevelopment of social and professional relationships and practices in and across workplaces and work spaces—"epistemic cultures," to revert to Knorr Cetina's (1999) term.

The physical collocation of scholars and researchers spawns trusting and "knowable communities" (Brown and Duguid 2000, 169), a phenomenon that has been observed in a variety of contexts. Von Hippel (1987) has described patterns of informal know-how trading between (competing) firms and Allen (1983) has identified a complementary phenomenon, namely, "collective invention." The importance of social networks to innovation is now widely recognized. In recent years, the significance of industry clusters (think of Silicon Valley or Silicon Glen) has been analyzed intensively in the business strategy and economic development literature, most notably by Porter (1998). The evidence shows that geographic concentration of interconnected companies, as well as the distribution of talent, or human intellectual capital (Florida 2002), can create local, regional, or national comparative advantage. Something similar appears to hold in academia, where the space of place (research sites, science parks) and the space of knowledge flows are tightly linked.

My colleague Katy Börner's (2003) goal is to help create a "Map of Science" screensaver in the tradition of ISI's prototypical *Atlas of Science*. Such an interface might get close to answering Otlet's question: "How could the intellectual domains be adequately, continuously mapped?" (quoted in Rayward 1992, 59). Börner envisages an automatically updated, 2-D, layered map based on data fed from all major publication, patent, and grant-funding databases. Users would be provided with a global overview of topic areas, data on the size and composition of fields, evidence of disciplinary emergence and merging, bird's-eye views of specific knowledge domains, highly influential papers, funding flows, and so forth. Such a development would be of practical value not only to career scientists but also science policy analysts, sociologists, and historians of science.

It is an attractive notion, but despite advances in information visualization, present generation toolsets and techniques cannot capture the biographical substrate of scholars' communicative practices, the complex and constantly shifting webs of collegiate relationships and "patronage networks" (Baber 2003, 96) that interleave with the purely cognitive ties to create the double helix of scholarly communication. Some degree of "social translucence" (Erickson et al. 2002) may be attainable within small, clearly bounded work groups, but it is hard to see how such a goal could be achieved in the unbounded world of scholarly communication. The social ties that bind scholars are often implicit, imperceptible, and evanescent. In exceptional cases it may be possible to generate "thick description" (Geertz 1973) of a kind that would faithfully describe the admixture of social, affective, and scholastic ties that exists within and between particular communities of scholars but, again, scalability remains a seemingly insurmountable obstacle. Indeed, it may well be that the "infrasociality" of science, to borrow Knorr Cetina's (1999, 13) term, is practicably indescribable.

And even if large-scale sociosemantic mapping of science were technically feasible, it seems unlikely that these forensic tools would reveal, to use White et al.'s (2004, 125) vivid metaphor, "an orgy of back scratching." But that is no reason for not trying to conceptualize and develop innovative tools (e.g., a "social search engine" [Barabási 2002, 39]) to harvest and graphically display sociocognitive influences within and across disciplines. Such visualizations would certainly bring us closer to generating reliable maps of scholarly trade routes and to understanding how social relations, forged in physical space, are reflected in the architecture of intellectual spaces.

REFERENCES

Allen, R. C. (1983). Collective invention. *Journal of Economic Behaviour and Organization*, 4, 1–24.

Allen, T. J. (1969). Information needs and uses. In: Cuadra, C. (Ed.). *Annual Review of Information Science and Technology,* 4. Chicago: Encyclopaedia Britannica, 3–29.

Baber, Z. (2003). The taming of science and technology studies. *Social Epistemology*, 17(2&3), 95–98.

Baldi, S. (1998). Normative versus social constructivist processes in the allocation of citations: A network-analytic model. *American Sociological Review*, 63, 829–846.

Barabási, A.-L. (2002). *Linked: How Everything is Linked to Everything Else and What It Means for Business, Science, and Everyday Life*. New York: Plume.

Boisot, M. H. (1995). *Information Space: A Framework for Learning in Organizations, Institutions and Culture*. London: Routledge.

Boisot, M. H. (1999). *Knowledge Assets: Securing Competitive Advantage in the Information Economy*. Oxford: Oxford University Press.

Bonto, J. P. (1996). *American Architects and Texts: A Computer-Aided Analysis of the Literature*. Cambridge, MA: MIT Press.

Borgman, C. L. and Furner, J. (2002). Scholarly communication and bibliometrics. In Cronin, B. (Ed.), *Annual Review of Information Science and Technology*, 36. Medford, N.J.: Information Today, 3–72.

Börner, K. (2003). Towards a cartographic map that shows the evolution of knowledge domains. Paper presented at the *99th Annual Meeting of the Association of*

American Geographers. Abstract available at http://convention.allacademic.com/aag2003/view_paper_info.html?pub_id=2580.

Börner, K., Chen, C., and Boyack, K. W. (2003). Visualizing knowledge domains. In: Cronin, B. (Ed.). *Annual Review of Information Science and Technology*, 37. Medford, NJ: Information Today, 179–255.

Boyd, D., Lee, H.-Y., Ramage, D., and Donath, J. (2002). Developing legible visualizations for online social spaces. In: *Proceedings of the Hawai'i International Conference on System Sciences*, January 7–10, Big Island, Hawaii.

Brown, J. S. and Duguid, P. (2000). *The Social Life of Information*. Boston, MA: Harvard Business School Press.

Burt, R. S. (1992). *Structural Holes: The Social Structure of Competition*. Cambridge, MA: Harvard University Press.

Castells, M. (2000). *The Rise of the Network Society*. Oxford: Blackwell.

Christensen, C. M. (2000). *The Innovator's Dilemma: When New Technologies Cause Great Firms to Fail*. Cambridge, MA: Harvard Business School Press.

Clark, A. (2003). *Natural-born Cyborgs*. Oxford: Oxford University Press.

Collins, R. (2003). Fuller, Kuhn, and the emergent attention space of reflexive studies of science. *Social Epistemology*, 17 (2&3), 147–152.

Crane, D. (1972). *Invisible Colleges: Diffusion of Knowledge in Scientific Communities*. Chicago: University of Chicago Press.

Cronin, B. (1984). *The Citation Process: The Role and Significance of Citations in Scientific Communication*. London: Taylor Graham.

Cronin, B. (1995). *The Scholar's Courtesy: The Role and Significance of Acknowledgement in the Primary Communication Process*. London: Taylor Graham.

Cronin, B. and Atkins, H. B. (Eds.) (2000). *The Web of Knowledge: A Festschrift in Honor of Eugene Garfield*. Medford, NJ: Information Today.

Cronin, B. and Shaw, D. (2001). Identity-creators and image-makers: using citation analysis and thick description to put authors in their place. *Scientometrics*, 54(1), 31–49.

Cronin, B. Shaw, D., and La Barre, K. (2003). A cast of thousands. Co-authorship and sub-authorship collaboration in the twentieth century as manifested in the scholarly literature of psychology and philosophy. *Journal of the American Society for Information Science and Technology*, 54(9), 855–871.

Cronin, B., Shaw, D., and La Barre, K. (2004). Visible, less visible, and invisible work: patterns of collaboration in twentieth century chemistry. *Journal of the American Society for Information Science and Technology*, 55(2), 160–168.

Davenport, E. (1993). *Risks and Rewards and Electronic Publishing: A Case Study of Information Science in the United Kingdom Using a Qualitative Methodology*. Glasgow, UK: University of Strathclyde, PhD thesis.

Davenport, E. and Cronin, B. (2000). The citation network as a prototype for representing trust in virtual environments. In: Cronin, B. and Atkins, H. B. (Eds.). *The Web of Knowledge: A Festschrift in Honor of Eugene Garfield.* Medford, NJ: Information Today, 517–534.

Davenport, E. and Cronin, B. (1990). Hypertext and the conduct of science. *Journal of Documentation*, 46(3), 175–192.

Dourish, P. (2001a). *Where the Action Is: The Foundations of Embodied Interaction.* Cambridge, MA: MIT Press.

Dourish, P. (2001b). Seeking a foundation for context-aware computing. *Human-Computer Interaction*, 16(2–4), 229–242.

Erickson, T., Halverson, C., Kellogg. W. A., Laff, M., and Wolf, T. (2002). Social translucence: designing social infrastructures that make collective activity visible. Available at www.pliant.org/personal/Tom_Erickson/Soc_ Infrastructures.html.

Etzkowitz, H. and Kemelgor, C. (1998). The role of research centres in the collectivisation of academic science. *Minerva*, 36(3), 271–288.

Florida, R. (2002). The economic geography of talent. *Annals of the Association of American Geographers*, 92(4), 743–755.

Furner, J. (2003). Little book, big book: Before and after *Little Science, Big Science*: A review article, Part I. *Journal of Librarianship and Information Science*, 35(2), 115–125.

Garfield, E. (1955). Citation indexes for science: A new dimension in documentation through association of ideas. *Science*, 122, 108–111.

Geertz, C. (1973). Thick description: Toward an interpretative theory of culture. In: *The Interpretation of Culture.* New York: Basic Books, 3–30.

Gladwell, M. (2002). *The Tipping Point: How Little Things Can Make a Big Difference.* New York: Back Pay Books.

Granovetter, M. (1973). The strength of weak ties. *American Journal of Sociology*, 78, 1360–1380.

Granovetter, M. (1983). The strength of weak ties: A network theory revisited. *Sociological Theory*, 1, 201–233.

Haythornthwaite, C. and Harger, C. (2005). The social worlds of the Web. In: Cronin, B. (Ed.). *Annual Review of Information Science and Technology*, 39. Medford, NJ: Information Today, 311–346.

Heimeriks, G. and van den Besselaar, P. (2002). *State of the Art in Bbliometrics and Webometrics.* Available at www.eicstes.org/EICSTES_PDF/ Deliverables/.

Hepworth, M. (1989). *Geography of the Information Economy.* London: Bellhaven Press.

Hutchins, E. (1995). *Cognition in the Wild.* Cambridge, MA. MIT Press.

Katz, J. S. (1994). Geographic proximity and scientific collaboration. *Scientometrics*, 31(1), 31–43.

King, D. W., Griffiths, J.-M., Sweet, E. A., and Wiederkehr, R. R. V. (1984). *A Study of the Value of Information and the Effect of Value of Intermediary Organizations, Timeliness of Services and Products, and Comprehensiveness of the EDB [Energy Data Base]*. Rockville, MD: King Research Inc. Report submitted to the Office of Scientific & Technical Information, U.S. Department of Energy, Oakridge, Tennessee.

Knorr Cetina, K. (1999). *Epistemic Cultures: How the Sciences Make Knowledge Work*. Cambridge, MA: Harvard University Press.

Koku, E., Nazer, N., and Wellman, B. (2001). Netting scholars: Online and offline. *American Behavioral Scientist*, 44(10), 1750–1772.

Kurtz, M. J., Eichorn, G., Accomazzi, A., Grant, C., Demleitner, M, Murray, S. S., Martimbeau, N., and Elwell, B. (2004). The bibliometric properties of article readership information. *Journal of the American Society for Information Science and Technology* (in press).

Latour, B. and Woolgar, S. (1979). *Laboratory Life: the Social Construction of Scientific Facts*. Beverly Hills, CA: Sage.

Lin, X., White, H. D., and Buzydlowski, J. (2003). Real-time author co-citation mapping for online searching. *Information Processing & Management*, 39(5), 689–706.

Mackinlay, J. D., Rao, R., and Card, S. K. (1995). An organic user interface for searching citation links. *Proceedings of the ACM Conference on Human Factors in Computing Systems*, 67–73.

Mählck, P. and Persson, O. (2000). Socio-bibliometric mapping of intra-departmental networks. *Scientometrics*, 49(1), 81–91.

Matthiessen, C. W., Schwarz, A. W., and Find, S. (2002). The top-level global research system, 1997–1999: Centers, networks and nodality. An analysis based on bibliometric indictors. *Urban Studies*, 39(5–6), 903–927.

Merton, R. K. (2000) (Eds.). On the Garfield input to the sociology of science: A retrospective collage. In: Cronin, B. and Atkins, H. B. (2000). *The Web of Knowledge: A Festschrift in Honor of Eugene Garfield*. Medford, NJ: Information Today, 435–448.

Merton, R. K. and Garfield, E. (1986). Foreword. In: Price, D. J. de Solla. *Little Science, Big Science . . . and Beyond*. New York: Columbia University Press, 1–4.

Odlyzko, A. (2002). The rapid evolution of scholarly communication. *Learned Publishing*, 15(1), 7–19.

Page, L., Brin, S., Motwani, R., and Winograd, T. (1998). *The PageRank Citation Ranking: Bringing Order to the Web*. Technical report, Stanford Digital Library Technologies Project. Stanford University.

Porter, M. E. (1998, November–December). Clusters: The new economics of competition. *Harvard Business Review*, 77–90.

Price, D. J. de Solla. (1965). Networks of scientific papers. *Science*, 149, 510–515.

Price, D. J. de Solla. (1970), Citation measures of hard science, soft science, technology, and nonscience. In: Nelson, C. E. and Pollock, D. K. (Eds.). *Communication Among Scientists and Engineers*. Lexington, MA: Heath, 3–22.

Rayward, W. B. (1992). Restructuring and mobilizing information in documents: A historical perspective. In: Vakkari, P. and Cronin, B. (Eds.). *Conceptions of Library and Information Science: Historical, Empirical and Theoretical Perspectives*. London: Taylor Graham, 50–68.

Rheinberger, H-J. (2003). "Discourses of circumstance": A note on the author in science. In: Biagioli, M. and Galison, P. (Eds.). *Scientific Authorship: Credit and Intellectual Property in Science*. New York: Routledge, 309–323.

Rowland, I. (1999). Patterns of author cocitation in information policy: Evidence of social, collaborative and cognitive structure. *Scientometrics*, 44(3), 533–546.

Salton, G., Wong, A., and Yang, C. S. (1975). A Vector Space Model for automatic indexing. *Communications of the ACM*, 18(11), 613–620.

Skupin, A. (2000). From metaphor to method: Cartographic perspectives on information visualization. *Proceedings of the IEEE Symposium on Information Visualization 2000 (InfoVis2000)*. Los Alamitos, CA: IEEE CS Press, 91–97.

Skupin, A. (2002). A cartographic approach to visualizing conference abstracts. *IEEE Computer Graphics and Applications*, 22(1), 50–58.

Small, H. (1973). Co-citation in the scientific literature: A new measure of the relationship between two documents. *Journal of the American Society for Information Science*, 24(4), 265–269.

Small, H. (1999). Visualizing science by citation mapping, *Journal of the American Society for Information Science*, 50(9), 799–813.

Snyder, H., Cronin, B., and Davenport, E. (1995). What's the use of citation? Citation analysis as a literature topic in selected disciplines of the social sciences. *Journal of Information Science*, 21(2), 75–85.

Suchman, L. (1987). *Plans and Situated Actions: The Problem of Human-Machine Communication*. Cambridge: Cambridge University Press.

Thelwall, M., Vaughan, L., and Bjöneborn, L. (2005). Webometrics. In: Cronin, B. (Ed.). *Annual Review of Information Science and Technology*. 39. Medford, NJ: Information Today, 81–135.

von Hippel, E. (1987). Cooperation between rivals: Informal know-how trading, *Research Policy*, 16, 291–302

White, H. D. (1990). Author co-citation analysis: overview and defense. In: C. L. Borgman (Ed.). *Scholarly Communication and Bibliometrics*. Newbury Park, CA: Sage, 84–106.

White, H. D. (2001). Authors as citers over time. *Journal of the American Society for Information Science and Technology*, 52(2), 87–108.

White, H. D. and McCain, K. W. (1997). Visualization of literatures. In: Williams, M. E. (Ed.). *Annual Review of Information Science and Technology*, 32. Medford, NJ: Information Today, 99–168.

White, H. D. and McCain, K. W. (1998). Visualizing a discipline: An author co-citation analysis of information science, 1972–1995. *Journal of the American Society for Information Science*, 49(4), 327–355.

White, H. D., Wellman, B., and Nazer, N. (2004). Does citation reflect social structure? Longitudinal evidence from the 'Globenet' interdisciplinary research group. *Journal of the American Society for Information Science and Technology*, 55(2), 111–126.

Williams, J. G., Sochats, K. M., and Morse, E. (1995). Visualization. In: Williams, M. E. (Ed.). *Annual Review of Information Science and Technology*, 30. Medford, NJ: Information Today, 161–207.

Wouters, P. (1993). Writing histories of scientometrics or what prescisely *is* scientometrics? Available from the author.

Wouters, P. (1999). *The Citation Culture*. University of Amsterdam. PhD thesis.

Wouters, P. (2000). Garfield as alchemist. In: Cronin, B. and Atkins, H. B. (Eds.). *The Web of Knowledge: A Festschrift in Honor of Eugene Garfield*. Medford, NJ: Information Today, 65–71.

Wouters, P. and Leydesdorff, L. (1994). Has Price's dream come true: Is scientometrcis a hard science? *Scientometrics*, 31(2), 193–222.

Ziman, J. (2000). *Real Science*. Cambridge: Cambridge University Press.

Zook, M. A. (2003). Underground globalization: Mapping the space of flows of the Internet adult industry. *Environment and Planning*, 35, 1261–1286.

Chapter Five

Intellectual Collaboration

We live in a rankings-besotted world. Vita building has become the academic equivalent of the workout. Even bodybuilders fret less about their vital statistics than academics. We routinely size up one another; evaluate curricula vitae, review dossiers, assess productivity, rate scholarship. We measure what we can (publications, citations, teaching, grant awards, honors, fellowships, invited presentations, etc.). Take the case of citation counting. The pros and cons of the approach have been aired exhaustively and probably don't need to be rehearsed here; suffice it to say that all citations—substantive, perfunctory, negative, or whatever—are treated as equal, though some are clearly more equal than others (a citation from a Nobel laureate presumably trumps one from a graduate student). The citing of one author by another is treated as a significant event in communication terms, and the more significant events recorded in favor of a particular author, the greater that author's presumed influence or prestige. As a consequence, citations have acquired convertible currency status in the academic marketplace; to be highly cited is to become hot property.

But citation counts are an incomplete register of influence or impact. There are other significant "events" that warrant attention, such as acknowledgment. How often do we find at the end of a journal article something like the following: "I am indebted to my colleagues X and Y who read early drafts of this paper and made many valuable comments and suggestions"? These statements can range from a one-line expression of gratitude to a research assistant to a paragraph acknowledging a multiplicity of material and/or cognitive influences from colleagues and mentors. Many articles carry compound acknowledgments, reflecting a mix of personal, moral, financial, technical, and conceptual support from institutions, agencies, coworkers, and peers. What are we to make of these testimonies to the contributions of a trusted assessor, mentor, or advisor? As things stand, being mentioned in the acknowledgment appended to a scholarly article (or doctoral dissertation or monograph) is to be consigned to limbo. In the academic reward system, the acknowledgment has no status; no cash-in value. While a glowing acknowledgment counts for naught, a negative citation still sets the register ringing. This is anomalous. We ought to take acknowledgments seriously and consider how citations and acknowledgments could be used conjointly to better understand the sociocognitive structures of disciplines and also consider whether they should play a part in the assessment of both individual and institutional research performance.

To develop an initial (and admittedly impressionistic) sense of prevailing practice some fifteen years ago, I analyzed all acknowledgment statements attached to research articles appearing in the *Journal of the American Society for Information Science* over a twenty-year period (Cronin 1991). Approximately half (444 articles) carried an acknowledgment of some kind; almost a quarter (244) made explicit reference to intellectual input (inspiration, guidance, critical comment, feedback, etc.) received from peers, mentors, or coworkers (420 names in all). Most of these individuals (85 percent) were mentioned once; only 25 received three or more mentions. This, of course, is just the kind of concentration we have come to expect in analyzing the distribution of citations; citation and acknowledgment stars are equally thin on the ground (Davis and Cronin 1993).

As the results of this exploratory survey showed, many researchers dutifully discharge their intellectual debts via acknowledgments—just as dutifully as they include relevant works in the reference lists and bibliographies that accompany their publications. Subsequent studies of other fields and disciplines, involving, in aggregate, tens of thousands of acknowledgments, have demonstrated incontrovertibly that acknowledgment is an important aspect of research and academic writing (Cronin 1995). This raises a number of issues and questions: (a) What is the degree of correlation between cited and acknowledged authors; are we dealing with identical, overlapping, or discrete populations? (b)

Is there a hidden population of influencers whose contributions as nurturers, stimulators, and mentors do not show up on publication and citation counts? (c) Why should these explicit expressions of intellectual debt not be treated and processed like citations? (d) Why does the humble acknowledgment not feature in the academic auditor's armamentarium; why does it not play a part in the tenure review process? (e) What are the norms of acknowledgment? (f) Could the protocols and practice of acknowledgment be standardized (the equivalent of a referencing house style) so that acknowledgment counting, like citation counting, would be amenable to automation? (g) What are the financial and practical constraints to producing, say, an acknowledgment index, comparable with, and complementary to, the *Science Citation Index*?

In many cases, the acknowledgment functions as a closet citation. It almost begs to be taken out, dusted down, and put to good use. Those who are concerned about equity in the context of the promotion and tenure process would do well to consider how existing indicators might be bolstered by the introduction of data on acknowledgment frequency. On the surface, there is a strong case for taking a closer look at the role played by acknowledgments in the primary communication process.

The old saying "strength in numbers" holds for research and scholarship. As Posner (2001, 540) has observed, "academic work increasingly is teamwork, just like industrial production." Numerous studies have documented the growth in interinstitutional, interdisciplinary and intersectoral scientific collaboration in the second half of the twentieth century, especially in "big science" (Bordons and Gómez 2000; Luukkonen, Persson, and Sivertsen 1992). The internationalization of science and scholarship is also irreversibly established: "Scientific research in our time is either global or ceases to be scientific" (Castells 2000, 125). Various theories have been put forward to explain the growth in collaboration, ranging from resource optimization and the progressive professionalization of science (Beaver and Rosen 1978; Eaton 1951) to functional explanations of collaborative behavior (Wray 2002) to "changes in the propensity of scholars to work outside their fields of specialization" (Laband and Tollison 2000, 639). According to Moody (2004, 235), "coauthorship is more likely in specialties that admit to an easier division of labor. Research methods seem particularly important, showing that quantitative work is more likely to be coauthored than nonquantitative work."

The most visible indicators of the trend to collaboration and also the increasing division of labor are national and international coauthorship rates, data on which can be mined from online bibliographic sources such as the citation indexes produced by the Institute for Scientific Information (Narin, Hamilton, and Olivastro 2001; Persson, Glänzel, and Danell 2003; Pontille 2003). Another, though less immediately accessible and less comprehensive, indicator would be the number of PIs and co-PIs on research grant applications to agencies such as the National Science Foundation. Coauthorship rates have risen across the board, but most dramatically in science, technology, and medicine (Cronin 2002, 560–61; Wray 2002, 159, 167). Collaboration is not a function of low professional rank or junior status; Nobel laureates collaborate intensively. According to Zuckerman (1977, 176), "the majority of investigations honored by Nobel awards have involved collaboration," and the trend increased during the course of the twentieth century reflecting "the secular shift to joint research in all the sciences." In other words, teamwork pays off, whatever your place in the pecking order. It is also worth noting that the rates of return to multiauthored papers, in terms of citations—an important form of symbolic capital in the academic reward system (Cronin and Shaw 2002; Gelman and Gibelman 1999)—are frequently greater than for single-authored papers (Glänzel 2002, 472).

In some fields, as we saw in chapter 3, the number of coauthors can sometimes reach the hundreds. In such cases, the scale and complexity of projects are invariably beyond the capabilities of an individual or a small group, requiring instead professionally managed teams of often globally distributed scientists supported by complex research infrastructures and significant levels of federal and/or corporate funding (by way of illustration, total R&D expenditures in the U.S. for 2003 were $284 billion [Frauenheim 2004]). Structural interdependence (among research groups, labs, and institutional partners) has become a fact of life in many scientific fields, supplanting "the privatized monastic rules of research" (Gilroy 2002, B20). Then there are disciplines such as astronomy and botany that have long relied on the observations and testimonies of teams of amateur observers—"citizen science" (van House 2002, 101). In short, the idea of the "lone wolf" scholar (Patel 1973, 92), though appealing, is largely anachronistic, not only in the natural and physical sciences but also the social sciences. That said, in the humanities, sole authorship continues to be the norm.

Coauthorship, though, is a not a complete measure of collaboration; not all forms of professional interaction are signaled so directly (Katz and Martin 1997). Historically, some types of collaboration—what Wray (2002, 152) has termed "collective but non-collaborative research"—have seen credit and responsibility vested publicly in one individual rather than in the research group or ensemble; one thinks of Shapin's (1995, chapter 4) description of Robert Boyle's work environment, and the Great Man's relationship with his technical assistants (see also Steiner 2003).

Some individuals whose names are included as coauthors may have contributed little or nothing to the work reported; what was earlier referred to as "honorific" authorship (Croll 1984; Rennie and Flanagin 1994). Others, who have made material contributions, may find that they are not mentioned at all or, at best, included in a paper's acknowledgments. These underappreciated individuals are often referred to as "ghost" authors (Rennie and Flanagin 1994). Such practices (gifting and ghosting) may not be pandemic, but in biomedicine, as we have seen earlier, they have become a significant and well-documented source of concern to researchers, publishers, and editors alike. Nor is the problem limited to biomedical publishing. A survey of the membership of the American Physical Society also revealed concerns about the ethics of scientific coauthorship (Tarnow 2002).

There is another measure which can be used alongside coauthorship in documenting and analyzing trends in scholarly collaboration—subauthorship collaboration, a term coined by Patel (1973), and employed subsequently by Heffner (1981). Subauthorship collaboration is made manifest in acknowledgments. In the twentieth century, the acknowledgment has become a constitutive paratextual feature (Genette 1997a, 1997b) of the academic journal article, as well as a potentially rich source of insight into the myriad forms of informal assistance and interaction which are otherwise invisibly inscribed in scholarly texts (Cronin 1995; Davenport and Cronin 2001). It furnishes documentary evidence of what Mullins (1973, 30) calls "trusted assessorship," and provides a means of exploring "the sociology of the invisible" (Star 1991; see also Strauss 1985).

As with citation, theoretical discussion of acknowledgment is often framed in terms of normatively grounded behaviors, a set of reciprocal practices for which a tacitly understood "governing etiquette" (Cronin

1995, 107) exists. Unlike citations, which point or link to other publicly accessible work, acknowledgments have limited "instrumental cognitive functionality," though—and to continue using Merton's (2000, 438) terminology—they do share with citations a number of "symbolic institutional functions." In fact, acknowledgments, rather like citations, "provide pellets of . . . peer recognition" (Merton 2000, 438). Combined, co- and subauthorship data could in theory provide us with the means to create a compound index of collaboration in research, as Patel (1973) suggested in his longitudinal study of such practices in four leading sociology journals. These two paratextual elements (byline and acknowledgment) constitute a cumulating record of sociocognitive connections and dependencies within and between scholarly discourse communities; but because acknowledgment data are not yet machine searchable and analyzable in the way that author and citation data are with ISI's citation databases, almost invariably they are ignored in sociometric analyses of scholarly communication and disciplinary interaction.

Relatively few analyses of acknowledgment practices, genres, and trends have been undertaken because of this very basic, practical limitation. There is thus a temptation to dismiss acknowledgments as little more than marginalia in the scientific literature; but as Hollander notes (2002, 63), "their significance emerges when they are aggregated." Precisely the same argument has been made—and convincingly made—in respect of citations (White 1990). Over the years, I have carried out a range of acknowledgment studies covering the journal literature of such fields as history, information science, psychology, and sociology (Cronin 1995). However, these and subsequent explorations (Cronin, Davenport, and Martinson 1997; Cronin and Shaw 1999) typically covered only five or ten years' worth of the literature, which makes it difficult to develop a reliable sense of whether and how the acknowledgment genre has coevolved with other facets of the scholarly article and, moreover, the discipline of which it is a byproduct. Hartley (2003) has looked at the comparative frequency with which single authors, pairs, and larger groups of authors acknowledge colleagues, and found that the fewer the number of authors, the greater the number of acknowledgments. Bazerman's (1984) description of the evolution of experimental articles in *Physical Review* (1893–1980)—though not dealing specifically with the phenomenon of acknowledgment—does at least help locate the convention in its larger historical and discursive context. He notes, for instance, how the practice emerged, faded, and resurfaced over the years in this one journal. On the other hand, Hyland's (2000) wide-ranging monographic analysis of the lexical and stylistic features of a number of scholarly textual genres—including research articles— makes no mention whatsoever of acknowledgments as either an established or emerging phenomenon in academic writing, though he has subsequently carried out a study of acknowledgments in graduate dissertations (Hyland and Tse 2004).

To date, there has been little systematic analysis of the evolution of the acknowledgment, and little discussion of its functional or symbolic significance in the political economy of academic research. A notable exception, methodologically speaking, was the study by Laband and Tollison (2000, 646). The authors estimated, inter alia, the monetary value of a colleague's helpful comments, in the process distinguishing between acknowledgments to individuals with the same institutional affiliation ("home thanks") and individuals affiliated with other institutions ("away thanks"). We (Cronin, Shaw, and La Barre 2003, 2004), for our part, attempted to chronicle the use of acknowledgments in twentieth-century scholarship by gathering, analyzing, and classifying more than 4,500 specimens covering a 100-year period. Our first study focused on two fields: psychology and philosophy. Data were gathered from two leading and persistent journals, *Psychological Review* and *Mind*. The former (American) is an example of the literature of the social sciences, while the latter (British) is an example from a humanities discipline. Both journals have been in existence continuously for more than 100 years and both are very highly regarded. By logging and examining every acknowledgment (including acknowledgments of funding support) in every issue in each of these journals over the course of the twentieth century, we wanted to observe whether and how this paratextual subgenre has developed, and also consider whether its stylistic form and ostensive purposes had coevolved with disciplinary practices and discourses.

Our results showed that the intensity of acknowledgment varied by discipline, reflecting differences in prevailing sociocognitive structures and work practices. We also gathered complementary data on rates of coauthorship to highlight the growing importance of collaboration and the increasing division of labor in contemporary research and scholarship. In a follow-up study, we (Cronin, Shaw, and La Barre 2004) tracked both the use of acknowledgments and also the growth of coauthorship in chemistry by analyzing and classifying over 2,000 specimens covering a 100-year period in the *Journal of the American Chemical Society* (*JACS*) (Stang 2003). We titled that paper "Visible, Less Visible, and Invisible Work," to convey the idea that scientific contributions are differentially etched in the scientific record. Authors and coauthors are clearly visible, which is hardly surprising since they belong to what Shapin (1995, 414–15) has termed the "core-set," and they are rewarded or, more accurately, reward themselves by having their names displayed prominently. Those who feature in acknowledgments are less immediately visible; as one knows, not everyone pays attention to the names embedded in the typical acknowledgment section attached to a scientific paper. And within this population there is a subgroup of acknowledgees comprising editors and referees, whose contributions to the published work may have to remain veiled in anonymity. By way of an aside, unlike what we observed among the psychologists and philosophers, chemists frequently acknowledge instrumental and technical support, suggesting that it is indeed possible to develop an acknowledgment profile for this "less visible" work. Finally, there are the ranks of support personnel and others whose routine, paid contributions may simply not be deemed worthy of formal mention, as the following extract from the ACS's *Handbook for Authors* seems to suggest (American Chemical Society 1978, 18–19):

> Contributions of persons, other than coauthors, who have added substantially to the work may be acknowledged in a separate paragraph at the end of the paper. Supporting staff including draftsmen, machinists, and secretaries should not *ordinarily* [italics added] be mentioned. Recognition of assistance should be stated as simply as possible. Nontechnical information such as grant numbers and sponsors also should be included in this paragraph, in a separate terminal paragraph, or in a footnote, as should mention of preliminary presentation of the material at a meeting.

JACS authors seem to be following this approach, with the result that some individuals' more mundane contributions to the collective enterprise (or work for hire) presumably remain invisible.

Our *JACS* data suggest that individual agency is, indeed, a fading phenomenon in the world of chemical research. In chemistry, the dominant model of scientific knowledge production is industrial rather than artisanal in character. It is helpful to think in terms of the "laboratory as collective workshop," to use Shapin's (1995, 367) phrase, a site in which labor is increasingly finely divided. And as in industrial organizations, it is presumably the case with modern chemistry and some other scientific disciplines that the progressive division of labor translates into greater operating efficiencies. Moreover, during the course of the twentieth century, chemists have clearly become dutiful acknowledgers of others' contributions to their endeavors, as have their counterparts in other disciplines. In part, this is a function of scale dynamics (chemical research is often synonymous with teamwork and large laboratories), though it is also quite possible that scientists—and scholars in general—have become more sensitized, in an age of litigiousness and rights management, to the need to acknowledge the diverse kinds of contributions that are needed to bring the fruits of their research into the public eye.

To summarize thus far: in some scientific fields, collaboration is a well-documented fact of life (Knorr Cetina 1999; Traweek 1992). But one does not need to resort to the usual suspects (high-energy physicists) to support this assertion. Numerous studies have shown, for example, the growth in levels of multiple authorship in recent decades—the most visible and concrete operationalization (along with records of research funding) of scientific collaboration. This is true not only for the physical and life sciences but also for the social sciences and, even, to some extent, the humanities. The explanation for this difference lies in the fact that the scale, experimental sophistication, and costliness of research in the physical and (harder) social sciences have grown much in postwar years (Beaver 2001). Growing specialization within disciplines is also a factor (Piette and Ross 1992). It would be tempting to conclude that philosophers continue to work alone while chemists and psychologists are, more than ever, team players. But the reality of scholarly publishing is rather more complex. Although coauthorship is a common reward for specialized collaboration, not all forms of collaboration result in formal recognition (Laudel 2001). In economics, for example, intellectual collaboration "takes two principal, not mutually exclusive, forms: formal coauthorship and 'informal' commentary on one another's work" (Laband and Tollison 2000, 633). That, as we shall see shortly, is true of most disciplines.

BOWLING ALONE . . . TOGETHER

Robert Putnam's (2000) hugely successful book *Bowling Alone* chronicles the reduction in the stock of social capital in contemporary American life. The titular metaphor is used to evoke the decline in various forms of collective activity and civic engagement across the nation and the corresponding growth of individualism. To be precise, more Americans are bowling than previously, yet fewer belong to bowling leagues. However, we should be careful not to overgeneralize from Putnam's findings. What holds for the public sphere does not necessarily apply to the world of science and scholarship. In fact, the reverse trend obtains. Cooperation and collaboration (bowling together) have become much more common during the twentieth century than individual research and investigation, such that the lone wolf scholar may soon feature on the endangered species list along with the bowling league.

The popular image of the humanities scholar as a solitary figure hunched over a desk in a research library or ferreting around in dusty archives is something of a caricature. Philologists and mathematicians, for instance, may not be especially inclined to coauthor or engage in, to use Marx's (1981, 199) terminology, "communal labour," but they still interact with their professional colleagues; and they draw upon the validated and accepted findings of prior generations of scholars, what Marx (1981, 199) termed "universal labour." At the risk of stating the obvious, sole authorship is not synonymous with intellectual solitary confinement. Philosophers, just like other scholars and scientists, are enmeshed in distributed, sociocognitive networks. Their ideas and insights do not usually appear in a scholarly monograph or journal of record without having been shaped, read, and critiqued by a handful of "trusted assessors"—those peers, both local and remote, who are routinely sent working papers and drafts for private comment prior to submission and editorial review (Chubin 1975)—and without having been given an airing in colloquia and symposia. Philosophers may not collaborate using formal structures, coalitions, and sophisticated tool sets, but they do interact, more or less intensively, with colleagues and resources both inside and well beyond their parent institutions. These often imperceptible forms of informal and frequent collaboration matter and, as I have shown, can be measured—albeit with some difficulty (Cronin 1995). With regard to mathematics, Zuccala (2004) has described the importance of invisible colleges in the dissemination of research findings and, more concretely, the role played by some of the major research institutes that have been established in the last few decades (e.g., the Fields Institute in Toronto and the Isaac Newton Institute in Cambridge) in creating opportunities for productive, collegial interaction.

Our (Cronin, Shaw, and La Barre 2003) diachronic study also showed that philosophers had become assiduous acknowledgers over the course of the twentieth century; in 1999, 94 percent of philosophy articles included an acknowledgment compared with zero in 1902. Many of these acknowledgments explicitly recognized the intellectual contributions of significant others to the stimulation and evolution of the work being reported. Obviously, the highly articulated collaboration associated with, say, a massive clinical trial is very different from the kinds of low-key, personal communications that take place between individual philosophers, but both disciplinary tribes nonetheless exhibit, and depend upon, collaborative behaviors. Formal collaboration is increasingly required for certain kinds of research where specialized division of labor is required (biomedicine and high-energy physics are assuredly the poster children of postmodern science) but that is not the case with philosophy (and some other disciplines). However, collaboration, as I use the term here, is meant to cover a broad spectrum of interactional possibilities, both formal and informal.

Publicly accessible evidence of scientists' and scholars' social embeddedness is nowhere more clearly and routinely inscribed than in acknowledgments, the inchoate ledger of trusted assessorship-in-action. In a way that citations cannot, acknowledgments bring us close to "orally transmitted influence, the interdependence of one person's thoughts and achievements with another's" (Wernick 1991, 169), even if sometimes (as with citing) the motivation (e.g., strategic coat-tailing) for acknowledging another may be "impure." Acknowledgments constitute a tapestry of essentially private interactions and interplays between scattered actors, artifacts, resources, and institutions, and they reflect a kaleidoscope of moral support and material influences (instrumental, financial, technical, editorial, etc.). For present purposes, I simply want to make three points: (1) Most recognized genres of academic writing include acknowledgments; (2) most acknowledgments include evidence of collaboration; and (3) much of the reported collaboration is both cognitive and distributive in character.

DISTRIBUTED COGNITION

The discussion so far brings us to the notion of distributed cognition (Hutchins 1995). There is a large and multistranded literature on distributed cognition (Halverson 2002) and related constructs (e.g., external cognition, Scaife and Rogers [1996]). According to Hutchins (1995, xiii) cognition is not "a solitary mental activity" but something "always situated in a complex sociocultural world." Cognition is much more than what goes on inside the head of a willing subject in an experimental setting; rather, "naturally situated cognition," to use Hutchins' term (xii), takes account of the material, cultural, and social environments in which thinking and action occur, and as such is more difficult to nail down than the actions of a subject in the controlled environment of, say, a usability lab. In our reasoning and problem solving, we instinctively and routinely mobilize a range of external resources and representations. Simply stated, we are inherently social thinkers, cognitively cantilevered into the *Umwelt*. Or, as Wittgenstein put it (quoted in Monk 1990, 316): "I think there is some truth in my idea that I really only think reproductively. I don't believe I have ever invented a line of thinking. I have always taken one over from someone else."

Permit me once again to personalize matters: in writing this chapter, I talked at length with my colleague Yvonne Rogers (an active researcher in the human-computer interaction and cognitive science communities), drew upon an array of textual resources (including the articles, papers, and books listed in the references that follow), made use of artifacts to hand (a pocket calculator, search engine, text editor, spellchecker, pen, and paper), and looked at a number of external representations (charts and diagrams). I did not sit down one Thursday morning and suddenly decide that it was time to compose a few words on the subject of academic writing and distributed cognition. First, the idea emerged slowly and unfocusedly from recent and not so recent exchanges with colleagues near and far. Second, my ability to build some kind of argument, linking what we know about authorial practices in the world of scholarly publishing with current thinking about distributed cognition in fields such as psychology, cognitive science, cultural anthropology, and education (e.g., Salomon 1997), was contingent on my having easy access to the thoughts of others, whether directly (via face-to-face conversation or e-mail exchange) or indirectly (via the published literature or archival e-mail/listserv postings).

This chapter is not just the byproduct of mental processes going on inside what Clark (2003, 8) calls "the ancient biological skinbag" (a/k/a my head); it testifies to the fact that I am part of "an extended cognitive system" (Clark 2003, 8)—a superbly distributed system that feeds and supports my cogitations. I may not have met Edwin Hutchins in person, but I have encountered him through the pages of his book, other publications, and his website; I did not have to come up with the idea of external cognition, as I was fortunate enough to collide with Yvonne Rogers; I did not have to go through the original data from our survey of twentieth-century coauthorship to calculate the various figures quoted above, but instead extracted them from tables—the use of spreadsheets being a good instantiation of "computational offloading" (Scaife and Rogers 1996); I did not have to recall all the background material I needed since I could swivel in my chair and consult my personal library—an example of an external memory system in use. And even if I could not remember the specific information I needed to write this part of the book, I at least knew where to find it (metaknowledge, in other words). As Norman (1988, 56) put it succinctly and provocatively, "[K]nowledge is still mostly in the world, not in the head."

In short, this chapter (this monograph, to use the language of Nersessian et al. [2003]), is the result of a cognitive partnership between me (the author) and a constellation of other social actors (scholars/authors with whom I interact either first- or secondhand), scattered resources, tools, and "mediating artifacts" (Halverson 2002, 252). It is not a solo effort, even if sole authored. I must, therefore, be careful not to implicitly endorse a notion of authorship that places undue "emphasis on originality and self-declaring creative genius" (Jaszi and Woodmansee 2003, 195). The text you are reading did not emerge abiogenically, as it were; rather, it came into being as a result of my interactions with a sociotechnical system comprising people, texts, "environmental props, aids and scaffoldings" (Clark 2003, 75). My cognitive interactions are differentially distributed in time and space. Even when we bowl alone, we are in a sense bowling with others. Further evidence for this assertion—at least as far as scholarly publication is concerned—comes from Hartley (2003), who found that significantly more single authors acknowledge the help of others than do pairs (of authors). That, of course, is not to say that one's ideas do not on occasion "emerge from within a solitary framework," to use a phrase suggested by an anonymous reviewer, but they are inevitably molded and matured as a result of our interactions with other social actors and documentary representations of their thoughts.

MANY HANDS

Evidence of the distributed cognition phenomenon is found in the bibliographies, reference lists, and acknowledgments that are essential elements of scholarly scripts. It is here that the traces of our cognitive interdependencies are recorded for those who care to look. Much current thinking in cognitive science and cognitive anthropology stresses "culture-in-the-world" (Shore 1996, 5). That is to say, the human mind is more than the sum of localized (interiorized) cognitions; our thoughts, capabilities, and actions are continuously shaped by, and coevolve with, elements of the external world and the cultural contexts in which we operate. The relevance of this for our understanding of academic authorship is not hard to grasp. The historical privileging of *the* author obscures the fact that almost all writing is in some sense collaborative, even if there happens to be a single individual who controls the pen or keyboard. Multiple identities and personalities are embossed in every single-authored text (as a quick perusal of the attached references and acknowledgments should make clear), but we persist with the fiction that scholars toil in isolation, that ideas emerge from within a solitary mind.

We should not lose sight of the fact that "writing is a social act" (Brodkey 1987, 54). The concept of "the presence of the multiple in the individual"—or "singularity"—was suggested in the course of an e-mail exchange (conducted while drafting this chapter) with Ron Day, who, displaying commendable critical reflexivity in respect of his own authorial and acknowledgment practices, wrote as follows (Day 2003, italics added):

> The importance of my name was less as an author of a text, and more as someone *coresponsible for its existence*, though greatly responsible for its existence with a certain form and with a certain argumentative structure, vocabulary, and conclusion. But, often in social science and science arti-

cles, a strong claim is still made through style, form, and institutional reward systems for authoritative presence ("authorial voice") which is
somewhat ironic given the murkiness in determining what constitutes an
author with "coauthored" (not the same as co-written, obviously) papers.

Day's use of the Heideggerian notion of "coresponsibility" nicely captures the idea of distributed cognition from an authorial perspective that
I am trying to convey. Indeed, the near synchronous online interaction
(virtual conversation) with Day itself vivifies the notion of distributed
cognition. What I was writing was being shaped in real time by an "absent voice," the kind of voice and input that would only become apparent to the reader via an acknowledgment. As Day (2003) put it in a subsequent email,

> the issue is one of making a text rather than of reproducing one's "own
> thoughts." Texts involve work and production, not the simple re-presen-
> tation of mental content. They are woven syntaxes of materials, and ideas
> or "concepts" are constructs that result from the play of vocabulary,
> rhetorical techniques, and the powers expressed by social forces, tradi-
> tions, habits, and not the least, also by desires for invention and for futures
> different than the ones commonly assumed possible.

Writing, in short, does not take place in a sociocognitive vacuum.

In biomedicine, research and writing are the result of (distributed) intellectual and logistical collaborations, particularly in the case of multicenter (multinational) clinical trials. As we saw, this has resulted in proposals to dispense with the idea of "authorship" altogether and replace it
with that of "contributorship" (Rennie, Yank, and Emanuel 1997). Such
an approach would allow us to record in explicit fashion the nature of the
different contributions made by each individual associated with the published research and thus remove, or reduce, the present levels of ambiguity (Davenport and Cronin 2001). Some of this already happens, in the
sense that the ostensibly second-order contributions of colleagues and
others are publicly specified in acknowledgments, but ironically, until
now at least, the reader has had no way of knowing what each of the
coauthors on the masthead actually contributed to the work reported in a
multiauthored paper. To put it otherwise, "[R]esearchers in large-scale
biomedicine projects tend to think of authorship in corporate terms, that
is, as stocks in a company that carry credit and responsibility in proportion to their share of the total value of the enterprise. To them their names

are, literally, their stocks" (Biagioli 2003, 263). This is a far cry from the Romantic conception of authorship and the worldview of most humanists and many others in the academy.

CONCLUSIONS

The collaborative nature of science and scholarship is revealed not only in the statistics of coauthorship (overt indicators) but also in the persistent growth of subauthorship collaboration (covert indicators) (Patel 1973). By examining the paratextual elements of scholarly publications (bibliographies and acknowledgments, in particular), we can develop a much finer sense of the extent to which scholars of all stripes depend on loosely coupled assemblies of significant others—"distributed decentralized coalitions" in Clark's (2003, 13) terminology—to develop and bring their ideas into the public gaze. The forms and intensity of collaboration and mutual dependence may vary from one discipline to the next (the material practices and information-seeking behaviors of electrical engineers and cultural anthropologists, to take two disciplines at random, are quite different) but, as our and other (Laband and Tollison 2000) data show, collaboration, whether explicitly announced via coauthorship or implied by the paratextual recognition of others' contributions, is on the increase across disciplines, albeit at differential rates. The individuals who provided input to not only this chapter but also the book of which it is part and whose names are to be found in the front matter are much more than bit players on the scholarly stage; they are living proof of the assertion that academic writing is a high-level instance of distributed cognition. Or, to use Moody's (2004, 215) words: "one's ideas are a function of position in a social setting, which is deeply structured by interaction patterns." They are my cognitive partners, even though they may not have known or, indeed, wished it. We, as authors, may be center stage, in the sense that it is our names that are up in lights and it is we who will be *the* authors of record, but many of our sole-authored publications emerge from a series of (essentially unplanned) interactions distributed over space and time. We may, some of us, bowl alone as far as the scorekeepers in the academic reward system are concerned, but we are routinely aided and abetted by sometimes unwitting, occasionally unseen, and not infrequently sidelined helpers.

REFERENCES

American Chemical Society (1978). *Handbook for Authors of Papers in American Chemical Society Publications*. Washington, DC. American Chemical Society.

Bazerman, C. (1984). Modern evolution of the experimental report in physics: Spectroscopic articles in *Physical Review*, 1893–1980. *Social Studies of Science*, 14(2), 163–196.

Beaver, D. de B. (2001). Reflections on scientific collaboration (and its study): Past, present and future. *Scientometrics*, 52(3), 365–377.

Beaver, D. de B. and Rosen, R. (1978). Studies in scientific collaboration. Part I. The professional origins of scientific co-authorship. *Scientometrics*, 1(1), 65–84.

Biagioli, M. (2003). Rights or rewards? Changing frameworks of scientific authorship. In: Biagioli, M. and Galison, P. (Eds.). *Scientific Authorship: Credit and Intellectual Property in Science*. New York: Routledge, 253–279.

Bordons, M. and Gómez, I. (2000). Collaboration networks in science. In: Cronin, B. and Atkins, H. B. (Eds.). *The Web of Knowledge: A Festschrift in Honor of Eugene Garfield*. Medford, NJ: Information Today, 197–213.

Brodkey, L. (1987). *Academic Writing as Social Practice*. Philadelphia: Temple University Press.

Castells, M. (2000). *The Rise of the Network Society*. Oxford: Blackwell.

Chubin, D. E. (1975). Trusted assessorship in science: A relation in need of data. *Social Studies of Science*, 5, 362–368.

Clark, A. (2003). *Natural-born Cyborgs*. Oxford: Oxford University Press.

Croll, R. P. (1984). The noncontributing author: An issue of credit and responsibility. *Perspectives in Biology and Medicine*, 27(3), 401–407.

Cronin, B. (1991). Let the credits roll: a review of the role played by mentors and trusted assessors in disciplinary formation. *Journal of Documentation*, 47(3), 227–239.

Cronin, B. (1995). *The Scholar's Courtesy: The Role of Acknowledgements in the Primary Communication Process*. London: Taylor Graham.

Cronin, B. (2002). Hyperauthorship: A postmodern perversion or evidence of a structural shift in scholarly communication practices? *Journal of the American Society for Information Science and Technology*, 52(7), 558–569.

Cronin, B. and Shaw, D. (1999). Citation, funding acknowledgement and author nationality relationships in four information science journals. *Journal of Documentation*, 55(4), 404–408.

Cronin, B. and Shaw, D. (2002). Banking (on) different forms of symbolic capital. *Journal of the American Society for Information Science and Technology*, 53(13), 1267–1270.

Cronin, B., Davenport, E., and Martinson, A. (1997). Women's studies: Bibliometric and content analysis of the formative years. *Journal of Documentation*, 53(2), 123–138.

Cronin, B., Shaw, D., and La Barre, K. (2003). A cast of thousands: Co-authorship and sub-authorship collaboration in the twentieth century as manifested in the scholarly journal literature of psychology and philosophy. *Journal of the American Society for Information Science and Technology*, 54(9), 855–871.

Cronin, B., Shaw, D., and La Barre, K. (2004). Visible, less visible, and invisible work: Patterns of collaboration in twentieth century chemistry. *Journal of the American Society for Information Science and Technology*, 52(2), 160–168.

Davenport, E. and Cronin, B. (2001). Who dunnit? Metatags and hyperauthorship. *Journal of the American Society for Information Science and Technology*, 52(9), 770–773.

Davis, C. H. and Cronin, B. (1993). Acknowledgments and intellectual indebtedness: A bibliometric conjecture. *Journal of the American Society for Information Science*, 44(10), 590–592.

Day, R. (2003, August 2, 8, 12). Personal communication.

Eaton, J. W. (1951). Social processes of professional teamwork. *American Sociological Review*, 16(5), 707–713.

Epstein, J. (1983). Dedications and acknowledgments. *Publishers Weekly*, 224, 43–45.

Frauenheim, E. (2004). Less-than-risky-business? Available at http://att.com .com/Less-than-risky+business%3F/2100-1008_3-5299245.html.

Garvey, W. D. and Griffith, B. C. (1972). Communication and information processing within scientific disciplines: Empirical findings for psychology. *Information Storage & Retrieval*, 8(3), 123–136.

Gelman, S. R. and Gibelman, M. (1999). A quest for citations? An analysis of and commentary on the trend toward multiple authorship. *Journal of Social Work Education*, 35(2), 203–213.

Genette, G. (1997a). *Palimpsests: Literature in the Second Degree*. Lincoln, NE: University of Nebraska Press.

Genette, G. (1997b). *Paratexts: Thresholds of Interpretation*. Cambridge: Cambridge University Press.

Gifford, R. (1988). Book dedications: A new measure of scholarly indebtedness. *Scholarly Publishing*, 19(4), 221–226.

Gilroy, P. (2002, July 26). Cultural studies and the crisis in Britain's universities. *Chronicle of Higher Education*, B20.

Glänzel, W. (2002). Coauthorship patterns and trends in the sciences (1980–1998): A bibliometric study with implications for database indexing and search strategies. *Library Trends*, 50(3), 461–473.

Halverson, C. A. (2002). Activity theory and distributed cognition: Or what does CSCW need to DO with theories? *Computer Supported Cooperative Work*, 11(1–2), 243–267.

Hartley, J. (2003). Single authors are not alone: Colleagues often help. *Journal of Scholarly Publishing*, 34(2), 108–113.

Heffner, A. G. (1981). Funded research, multiple authorship, and subauthorship collaboration in four disciplines. *Scientometrics*, 3(1), 5–12.

Hollander, P. (2002). Acknowledgments: An academic ritual. *Academic Questions*, 15(1), 63–76.

Hutchins, E. (1995). *Cognition in the Wild*. Cambridge, MA: MIT Press.

Hyland, K. (2000). *Disciplinary Discourses: Social Interactions in Academic Writing*. London: Longman.

Hyland, K. and Tse, P. (2004). "I would like to thank my supervisor." Acknowledgements in graduate dissertations. *International Journal of Applied Linguistics*, 14(2), 259–275.

Jaszi, P. and Woodmansee, M. (2003). Beyond authorship: Refiguring rights in traditional culture and bioknowledge. In: Biagioli, M. and Galison, P. (Eds.). *Scientific Authorship: Credit and Intellectual Property in Science*. New York: Routledge, 195–223.

Katz, J. S. and Martin, B. R. (1997). What is research collaboration? *Research Policy*, 26(1), 1–18.

Knorr Cetina, K. (1999). *Epistemic Cultures: How the Sciences Make Knowledge*. Cambridge, MA: Harvard University Press.

Laband, D. N. and Tollison, R. D. (2000). Intellectual collaboration. *Journal of Political Economy*, 108(3), 632–662.

Laudel, G. (2001). What do we measure by co-authorships? *Proceedings of the 8th International Conference on Scientometrics and Informetrics*. Sydney, Australia, Vol. 1, 369–384.

Luukkonen, T., Persson, O., and Sivertsen, G. (1992). Understanding patterns of international scientific collaboration. *Science, Technology, & Human Values*, 17(1), 101–126.

Marx, K. (1981). *Capital: A Critique of Political Economy* (Vol. 3) (D. Fernbach, Trans.). Harmondsworth, UK: Penguin Books.

Merton, R. K. (2000). On the Garfield input to the sociology of science: A retrospective collage. In: Cronin, B. and Atkins, H. B. (Eds.). *The Web of Knowledge: A Festschrift in Honor of Eugene Garfield*. Medford, NJ: Information Today, 435–448.

Monk, R. (1990). *Ludwig Wittgenstein: The Duty of Genius*. New York: Penguin.

Moody, J. (2004). The structure of a social science collaboration network: Disciplinary cohesion from 1963 to 1999. *American Sociological Review*, 69, 213–238.

Mullins, N. C. (1973). *Theories and Theory Groups in Contemporary American Sociology*. London: Harper & Row.

Narin, F., Hamilton, K. S., and Olivastro, D. (2001). The development of science indicators in the United States. In: Cronin, B. and Atkins, H. B. (Eds.). *The Web of Knowledge: A Festschrift in Honor of Eugene Garfield*. Medford, NJ: Information Today, 337–360.

Nersessian, N. J., Kurz-Milcke, E., Newstetter, W. C., and Davies, J. (2003). Research laboratories as evolving distributed cognitive systems. In: *Proceedings of the Twenty-Fifth Annual Conference of the Cognitive Science Society*, Boston, July, 2003. Available at www.cc.gatech.edu/aimosaic/faculty/nersessian/.

Norman, D. A. (1988). *The Psychology of Everyday Things*. New York: Basic Books.

Patel, N. (1973). Collaboration in the professional growth of American sociology. *Social Science Information*, 12(6), 77–92.

Persson, O., Glänzel, W., and Danell. R. (2003). Inflationary bibliometric values: The role of scientific collaboration and the need for relative indicators in evaluative studies. *Proceedings of the 9th International Conference on Scientometrics and Informetrics, Beijing*, 411–420.

Piette, M. J. and Ross, K. L. (1992). An analysis of the determinants of co-authorship in economics. *Journal of Economic Education*, 23, 277–283.

Pontille, D. (2003). Authorship practices and institutional contexts in sociology: Elements for a comparison of the United States and France. *Science, Technology, & Human Values*, 28(2), 217–243.

Posner, R. A. (2001). *Public Intellectuals: A Study of Decline*. Cambridge, MA: Harvard University Press.

Putnam, R. D. (2000). *Bowling Alone: The Collapse and Revival of American Community*. New York: Simon & Schuster.

Rennie, D. and Flanagin, A. (1994). Authorship! Authorship! Guests, ghosts, grafters, and the two-sided coin. *Journal of the American Medical Association*, 264(6), 1857.

Rennie, D., Yank, V., and Emanuel, L. (1997). When authorship fails: A proposal to make contributors accountable. *Journal of the American Medical Association*, 287, 579–585.

Salomon, G. (Ed.). (1997). *Distributed Cognitions: Psychological and Educational Considerations*. Cambridge: Cambridge University Press.

Scaife, M. and Rogers, Y. (1996). External cognition: how do graphical representations work? *International Journal of Human-Computer Studies*, 45, 185–213.

Shapin, S. (1995). *A Social History of Truth: Civility and Science in Seventeenth-Century England*. Chicago: University of Chicago Press.

Shore, B. (1996). *Culture in Mind: Cognition, Culture, and the Problem of Meaning*. Oxford: Oxford University Press.

Stang, P. J. (2003). Editorial. 124 years of publishing original and primary chemical research: 135,149 publications, 573,453 pages, and a century of excellence. *Journal of the American Chemical Society*, 125, 1–8.

Star, S. L. (1991). The sociology of the invisible: The primacy of work in the writings of Anselm Strauss. In: Maines, D. (Ed.). *Social Organization and Social Process: Essays in Honor of Anselm Strauss*. Hawthorne, NY: Aldine de Gruyter, 1991, 265–283.

Steiner, G. (2003). *Lessons of the Masters*. Cambridge, MA: Harvard University Press.

Strauss, A. (1985). Work and the division of labour. *The Sociological Quarterly*, 26(1), 1–19.

Tarnow, E. (2002). Coauthorship in physics. *Science and Engineering Ethics*, 8(2), 175–190.

Traweek, S. (1992). *Beamtimes and Lifetimes: The World of High Energy Physics*. Cambridge, MA: Harvard University Press.

van House, N. A. (2002). Digital libraries and practices of trust: Networked biodiversity information. *Social Epistemology*, 16(1), 99–114.

Wernick, A. (1991). *Promotional Culture*. Newbury Park, CA: Sage.

White, H. D. (1990). Author co-citation analysis: Overview and defense. In: Borgman, C. L. (Ed.). *Scholarly Communication and Bibliometrics*. Newbury Park, CA: Sage, 84–106.

Wray, K. B. (2002). The epistemic significance of collaborative research. *Philosophy of Science*, 69(1), 150–168.

Zuccala, A. (2004). *Revisiting the Invisible College: A Case Study of the Intellectual Structure and Social Process of Singularity Theory Research in Mathematics*. PhD dissertation. University of Toronto.

Zuckerman, H. (1977). *Scientific Elite: Nobel Laureates in the United States*. New York: Free Press.

Chapter Six

The Reward System

In May 2002, Stephen Greenblatt, then president of the Modern Language Association (MLA), sent a "special letter" (*Call for Action on Problems in Scholarly Publishing*) to all members of the Association (Greenblatt 2002). The letter outlined the pressures faced by junior faculty in some humanities disciplines who are expected to produce a full-length scholarly book (sometimes two) published by a reputable press for tenure and promotion, and this at a time when university presses are struggling to maintain their minority-appeal academic lists. Greenblatt correctly characterized the problem facing the academy as "systemic, structural, and at base economic," and put forward two concrete proposals to alleviate the situation. The first encouraged university tenure and promotion (T&P) committees to show greater willingness to accept alternative forms of evidence of scholarly output, such as collections of significant journal articles. The second considered how junior scholars might be provided with a first-book subvention to help them break into print; this proposal would essentially preserve the status quo, while alleviating the publishing logjam.

Greenblatt's much discussed missive was followed by the publication of a report, *The Future of Scholarly Publishing*, produced by an MLA ad hoc committee, which described the nature of the "widely perceived crisis in scholarly publishing" (Ryan et al. 2002, 172). The report can be summarized as follows: library budgets for monographs are declining, as are subsidies for university presses; rising expectations for tenure and promotion result in ever more submissions to presses; additionally, it appears to be the case that presses are publishing fewer specialized studies, at least in foreign language fields; increasing competition among faculty candidates for entry-level positions has allowed hiring institutions to raise the performance bar, forcing young scholars to publish more and earlier. In short, junior scholars face rising standards at a time of overcapacity in a highly competitive marketplace; in this regard, see also Regier's (2003) analysis of university presses, Davidson's (2003) recommendations for reform and Monaghan (2004) on the debate about subventions for authors.

Others have addressed the perceived crisis in humanities publishing. Lindsay Waters (2001), executive editor for the humanities at the Harvard University Press, wrote a feisty essay on the "tyranny of the monograph" in the *Chronicle of Higher Education*, in which he lamented the "exaggerated emphasis on the publication of books" and the emergence of a culture that "is demanding more and more publications from scholars at a younger and younger age." The result of these dual pressures, he maintains, is "a lot of Mickey Mouse work." His views on the eclipse of scholarship in the humanities have since been packaged in the form of a brief monograph, *Enemies of Promise* (Waters 2004). Elsewhere, the call for reform could also be heard; in the editor's column of the *Proceedings of the Modern Language Association (PMLA)*, Carlos Alonso (2003) examined the merits and drawbacks of the proposals, fleshing out the mechanics of a possible subvention scheme and the role professional bodies such as the MLA might play in the future.

On balance, Alonso seems to believe that the book should persist as the central element in tenure decisions because crafting a series of articles does not quite compare with the conceiving and writing of a full-length book. Moreover, he adds (2003, 221), the reader of a book "receives its argument in a condensed and organic manner that a series of related articles published seriatim can never hope to match." Whatever the merits of Alonso's argument, it is worth noting that the production of a book-length manuscript is not necessarily synonymous with its publication by a university press; production and publication can be decoupled, just as peer review and journal publication can be decoupled.

Irrespective of the relative attractiveness of Greenblatt's two proposals, there is a larger set of issues relating to institutional culture and faculty governance that needs to be taken into account, at least as far as the North American context is concerned. A skeptical Lewis (2002,

1224) wonders how "a vast and diverse system with little central regulation can be restructured so as to induce colleges and universities to adopt appropriate, institution-specific criteria for granting tenure and for understanding the obligations of tenured faculty members." And well he might; in academia, as Kennedy (1994, 97) has observed, "sunset is an hour that almost never comes" and, he could have added, certainly not at the same hour across the nation's many campuses. Even well-intentioned and largely well-received proposals for change to the status quo in departments of languages and literature, such as those promulgated by Greenblatt, can expect a guarded reception. As March, Schulz, and Zhou (2000, 48) noted in their diachronic analysis of rule making at Stanford University (including rules related to tenure, as it happens), "[t]oday's rules are often the solution to yesterday's problems." They go on to say that "[r]ule changes are often preceded by incidents, crises, or controversies." That observation seems pertinent in the light of Greenblatt's remarks. Although there is widespread recognition of the problems facing the humanities (Waters 2004), the role of senior faculty in resisting change receives little overt attention. One is reminded of Lohmann's (2002, 15) assertion that "it is the tenured faculty, above all, who are the fundamental source of intellectual and structural ossification."

As we saw in chapter 2, much has been written about the putative "crisis in scholarly publication," the effects of which are being felt well beyond the relatively cloistered world of the humanities. The established order is being challenged on many sides. The ways in which scholars communicate and archive the results of their research today are quite different from even a decade or so ago. New modes of electronic publishing and communication (from self-posting/self-archiving to blogging) are emerging, new forms of collaboration are establishing themselves and new approaches to peer review are being trialed (Andersen 2003). However, there is little evidence or prospect of standardization, either technologically or behaviorally, across disciplines. Within the humanities, change of some kind seems likely for the reasons sketched above, but to what extent will the primacy of the book be challenged; to what extent, and how, will digital scholarship be embraced by T&P committees (Andersen 2003; Cronin and Overfelt 1995)? That said, there is considerable inertia in the present system. The director of graduate studies in Harvard's English department, Barbara

Lewalski, does not see her institution lessening its commitment to the book as the sine qua non of tenure: "The kind of experience and training that's involved in producing a sustained piece of research and writing is rather different from the kind of training and experience that's gleaned from doing a 25-30-page journal article. I would be sorry to see us give [the book] over out of, as it were, marketplace pressures" (quoted in Ruark and Montell 2002). This sentiment is echoed by Peter Manning, chair of the English department at SUNY Stony Brook: "I would still say that if you go to an institution like Stony Brook or Berkeley or Michigan or Wisconsin, you had better have a book manuscript accepted by the time you come up for tenure. There are always going to be exceptions, but you would be really foolish to count on being one of them" (again, quoted in Ruark and Montell 2002).

Tenure and promotion guidelines determine who will be hired (retained, more correctly) and fired. However, this blunt binary is often buried in a morass of legalistic prose. An outside observer would surely be puzzled by the prolixity of the rules associated with recruiting and promoting university faculty. One might be forgiven for thinking that senior professors would be able to tell good work from bad, bright sparks from also-rans. But apparently not; to evaluate academic performance is to walk in a minefield. "Lawyers," it has been said (David and Spence 2003, iv), "devise contracts by imagining everything that could go wrong with a particular relationship and either providing solutions to those possible disputes, or mechanisms for resolving them if and when they arise." The word "faculty" could easily be substituted here for "lawyers." At times, attempts to operationalize "excellent," "outstanding" and similar adjectives are worthy of medieval schoolmen, not to mention contemporary attorneys—as Readings (1996, 24) implied in a chapter entitled "The Idea of Excellence" in his book, *The University in Ruins*, much quoted in this context, and most recently by the sitting president of the MLA, Robert Scholes.

> In this culture of "excellence," any shift from a book to articles as a standard for promotion is perceived as a reduction in standards. . . . [t]he question—the great question—that emerges from all this is whether we humanists can do anything to humanize this situation, to give substance to the empty signifier excellence, and to make the processes of intellectual growth and professional development more harmonious. (Scholes 2004, 3)

In 2003, we (Cronin and La Barre 2004) conducted a survey of language and literature departments in all U.S. Research I universities to explore guidelines relating to promotion and tenure. We found various definitions of scholarly quantity, quality, and impact—the holy trinity of performance indicators—in the corpus we examined. The choice of language was at times intentionally vague, a fact noted, too, by Kling and Spector (2003, 91–94); occasionally, the desire to eliminate any possible ambiguity results in phrasing worthy of a legal contract. Sometimes there is a tacit or even explicit admission that quality is ineffable; at other times, we found very precise output measures (variously four, five, six, or seven scholarly articles in lieu of a book; an average of one published article per year) being laid down. None of this is altogether new or surprising, but when one is exposed to so many guidelines the cumulative effect is eye opening. It is tempting to argue that since university departments of literature and language have more in common than not, they should be able to play down local differences (in norms, history, etc.) and concentrate on forging common understandings and approaches. Moreover, universities draw upon and contribute to a common pool of human intellectual capital, both for making academic appointments and for selecting external reviewers. Even so, any temptation to homogenize guidelines and procedures needs to be considered carefully. Given that our sample included, by way of illustration, Stanford and Old Dominion, it would be naïve to imagine that interinstitutional differences might not have a bearing on the definition and application of scholastic standards. After all, rules and procedures are institutionally embedded, and institutions differ appreciably.

The United States has an unremittingly litigious culture, and that, sadly, is now reflected in the way the academy conducts its affairs. Challenges to negative tenure and promotion decisions, whether for alleged process infringements or perceived bias (personal or standpoint-related), are commonplace, as a quick perusal of the *Chronicle of Higher Education* will make clear. Even though federal judges have fought shy of interfering in this aspect of a university's business (see www.harvard-magazine.com/on-line/090364.html), the stakes are nonetheless high in terms of potential damage to both individuals' and institutions' reputations. March et al. (2000, 51) note that "organizational rules are closely related to problems and crises in organizations," and our survey would seem to bear out their observation. The soi-disant crisis in scholarly publishing, coupled with the increasingly adversarial nature of the tenure process, has exacerbated a situation that was already disposed to breed rules and meta-rules—"rule density," to use March et al.'s (2000, 64) term. And, over time, both faculty and administrators have become more familiar with, and adept at, contesting, circumnavigating, and exploiting the system of rules to their respective advantage—legitimately or otherwise. Fear of failure has spawned fear of litigation, and fear of litigation has, in turn, spawned language worthy of parody.

These contextual observations aside, we came away from our survey with the clear sense that the granting of tenure in humanities departments still requires the production of a research monograph published by a reputable press. Sole authorship is also expected; the documents we examined are virtually silent on the issue of collaboration and coauthorship, unlike many other disciplines where single-authored articles are now almost anachronistic. Equivalence may, in theory, be attained by publishing, for example, a solid collection of peer-reviewed journal articles but—and here we are generalizing—the persistent and ubiquitous expectation is that junior scholars will have completed at least one book before moving up the ranks. (The same expectation holds for subsequent promotion to full professor.) In short, we have to say that the available evidence largely supports the core contention of inflexibility made by Greenblatt in his letter and also by Scholes in his president's column to the MLA membership, a finding echoed to some degree in a follow-up survey of history, English, and anthropology departments by Estabrook (2003). On the other hand, we found no clear evidence that written T&P criteria require a second book, unless the candidate's first book is deemed to be a straightforward repackaging of his or her PhD dissertation. A few institutions acknowledged the acceptability of online, electronic, and digital forms of scholarly production, but most were content to stress the importance of a candidate's work being subjected to peer review while remaining silent on the matter of medium. Overall, new modes of scholarly production and distribution received hardly any explicit attention.

Of course, our study analyzed mere words. What is needed is a complementary, phenomenological investigation to explore what actually goes on in the murky world of academic preferment. That would require gaining access to confidential discussions in T&P committees and also talking to both junior and senior faculty about their impressions and beliefs relating to the workings of the system. Not an easy task, it must be conceded, but one worth undertaking in the interests of both equity and economics.

RATES OF RETURN TO CITATION

Quantity may still count for much in the academic reward system, but the intrinsic quality and disciplinary impact of one's scholarship are increasingly important variables, and not just in the humanities. Publication alone is unlikely to be sufficient to justify tenure or promotion at most major research institutions. Performance review committees will also be looking for evidence of impact, sometimes, albeit incorrectly, seen as a synonym for quality; in short, has a scholar's published output been critically appraised and drawn upon by his disciplinary peers? Impact can, of course, be demonstrated in a variety of ways, but perhaps the easiest and (in the hard sciences, at least) the most commonly used metric is citation. Has one's work been cited in the literature, how often, by whom, and in which contexts? And how does one's citation count compare with others in one's field at the same stage of their careers? Citation scores matter, not just symbolically, but economically in academia. Research, as I discovered following an extensive review of the literature, shows that the rates of return to citation are substantial in the context of the academic reward system. Since quantitative performance measures, notably publication and citation counts, are associated strongly with life-cycle remuneration and career mobility, both could, assuming due diligence, be utilized in research assessment exercises. In the remainder of this chapter, I bring together empirical findings from the noninteracting research literatures of economics and information science to outline the case for using citation rates as a (more or less) valid and (relatively) cost-effective proxy for quality in certain disciplines.

In 1986 the University Grants Committee (UGC) conducted the first-ever research selectivity exercise in British university departments. Since then, the Research Assessment Exercises (RAEs), as they are now known, have become an inescapable feature of academic life in the United Kingdom. If nothing else, these rolling reviews of university departments, conducted by the UGC's successor bodies, the various Higher

Education Funding Councils (see www.rae.ac.uk for an overview of recent and planned future research assessment exercises and the methods to be used), have helped focus critical attention on the validity and reliability of the different measures used by, and available to, members of the review groups responsible for producing the quality rankings for the seventy-two subject areas covered by the exercise (Hodges et al. 1996). Despite repeated refinements to the criteria and conduct of RAEs, there is residual skepticism of attempts to define and measure the quality of research and scholarship, not least when it is proposed that evaluation be based on quantitative, in particular bibliometric, indicators (Anderson 1991). There have been signs for some time, however, that the United Kingdom is becoming more outcomes focused; the Economic and Social Research Council (ESRC) developed a Research Activity and Publications Information Database (RAPID), which contains summary information on all research funded by the council (Burnhill and Tubby-Hille 1994), and the Wellcome Trust established a commercial service for tracking acknowledgments in biomedical journal articles to research funding agencies (Jeschin, Dawson, and Anderson 1995). In the United States, impressionistic evidence suggests that publication and citation metrics are more readily accepted and more liberally applied, especially in the case of promotion and tenure decisions, or when program comparisons are being undertaken by funding bodies (Jaffe 2002).

The following example illustrates the greater tolerance of evaluative bibliometrics in the North American context. In a legal action initiated by a female assistant professor of biology who had been denied tenure at Vassar College, the plaintiff's lawyer brought forward as evidence of discrimination the fact that her untenured client had a higher citation count than some tenured male staff in the same department (Leatherman 1995). Although the female candidate's case was overturned subsequently on appeal (in part, and ironically, as a result of errors in the citation data submitted as evidence), the legal admissability and potential courtroom impact of citations are worthy of note. While this is a somewhat atypical, but not unique (Garfield 1979, 71–72), case, it should be noted that many candidates seeking promotion and tenure, not to mention the faculty committee members who judge their dossiers, routinely factor in citation data, though the relative weights attached to citations versus, say, type or place of publication may well vary from institution to institution.

CITATIONS AS PROXIES

One of my goals here is to establish a link between two logically related but mutually isolated or noninteractive literatures—to use Swanson's (1989, 432) terminology—dealing with the validity of citations in the context of research performance assessment. These two literature sets, which ought to be mutually reinforcing, are in fact separated by disciplinary boundaries and, to some extent at least, by time—an instance of a "structural hole" (Burt 1992), to use a phrase introduced earlier. One is a fairly small set of information science articles dealing with citations and departmental ratings; the other comprises a series of studies undertaken by (in the main) economists during the 1970s and 1980s to determine the direct career and monetary benefits of various categories of scholarly activity, such as teaching, research, publication, and administration. These studies also take particular account of the contribution made by citations to a scholar's life-cycle remuneration and career mobility. To summarize in advance: the combined findings from the economics and information science literature are apparently sufficiently robust to warrant using aggregate citation data in certain subject fields as a reasonable proxy for research quality in those same fields.

Although Garfield (1988) has produced a brief meta-analysis of research dealing with the relative rates of return from different kinds of scholarly output, very few information scientists seem to have tapped into this highly relevant subset of the literature on wage determination. Take the case of Diamond's (1986a) article, "What Is a Citation Worth?" which was published originally in the *Journal of Human Resources*. Given the persistent debate as to the validity and utility of citations as indicators of quality, how is it that such an explicitly titled article of substance has been so little quoted? As we shall see, the work of Diamond and other economists who have contributed to this area for almost three decades is noticeable by its continuing absence from both the U.K. and U.S. information science and higher-education policy literatures.

It is not difficult to find evidence to support the view that research assessments based solely on peer review may produce less than perfect results. Nor is it difficult to find evidence suggesting that reliance on publication and/or citation indicators alone may produce less than perfect results. Not surprisingly, many commentators favor what Martin and

Irvine (1983) have termed the converging partial indicators approach —
a judicious combination of hard and soft performance measures. In addition, special pleading is often heard on behalf of certain fields, academic activities, or modes of scholarly inquiry to which bibliometric
evaluation is inimical. Collini's (1989, 13) rhetorical question captures
the spirit: "Is it irrelevant that our literary critics write poems and nov-
els, that our historians develop television documentaries, that our economists challenge official statistics in the weeklies, and so on?" Ultimately, however, policymakers and institutional evaluators have to
make trade-offs between comprehensiveness and cost-effectiveness. In
practice, partial as opposed to perfect measures may have to suffice.
This, in turn, raises the question as to which indicators in which disciplines can best act as proxies for more elaborate combinatorial metrics
in the assessment of research performance, without inducing serious
distortions into the resultant rankings (see Warner [2000] for a methodological critique of citation analysis in the context of the RAE).

Two early papers by Oppenheim (1995) and Seng and Willett (1995)
sought explicitly to compare departmental rankings derived from bibliometric data with the official ratings produced by the 1992 RAE for
the subject area "Library and information management." Although they
used slightly different methods, both studies had a great deal in common — not least the conclusions they reached. The authors' shared point
of departure was straightforward: the RAE exercise concentrated on the
publication records of faculty, but took no account of citations attracted
by those publications. Both papers opened with a brief review of research into the validity of citations as measures of research quality, and
then proceeded to compare the citation profiles of the university departments in question; Oppenheim covered sixteen, while Seng and
Willett restricted their sample to seven institutions. Both studies computed Spearman rank order correlations between the RAE rankings and
a variety of bibliometrically based ratings. Oppenheim's (1995, 23) coefficients of correlation (between rank and number of citations/number
of citations per faculty member) were 0.81 and 0.82, respectively. Seng
and Willett's (1995, 70) coefficients were 0.873 (number of publications); 0.96 (total number of citations); 0.546 (mean number of publications per faculty member); 0.946 (mean number of citations per faculty member). Seng and Willett's study (1995, 71) demonstrates that as
far as the library and information science field in the United Kingdom

is concerned, there is "a strong statistical correlation between a department's citation counts and its RAE ranking." Oppenheim (1995, 25) comes to a similar conclusion, albeit via a slightly different route. He notes that the judgment made by the panel of eight relevant experts in the RAE exercise correlates strongly with citation counts, which, of course, only allows him to say that "there is a high association between rankings computed by citation counting and rankings by peer group assessment." However, he goes on to ask why, given virtually identical outcomes, the RAE "is carried out in such a laborious manner and why citation counting is not used instead?" Naturally, what holds for the library and information management subject area may not hold for others but, as he says, that is a hypothesis worth testing, in view of the potential savings and efficiency gains that could result.

Not all the results have been so persuasive. An earlier attempt to compare rankings produced by peer review with departmental citation counts was carried out by Zhu and Meadows (1991) using data generated in the course of the 1987 assessment exercise, initiated by the now defunct UFC (University Funding Committee). Zhu and Meadows selected two chemistry departments for detailed scrutiny, if only because chemistry departments tend to be among the highest producers of scholarly articles, thereby ensuring a solid statistical base for bibliometric analyses. The departments were of similar size (in terms of faculty) and located in equivalently sized universities. One had received a UFC grading of two and the other a five, on a scale of one to five, where five implied distinction.

However, their results were less convincing than those obtained by either Oppenheim or Seng and Willett. Despite the perceived rating differences, the number of publications per faculty member and the number of citations per publication showed no clear-cut differences between the two chemistry departments, leading Zhu and Meadows (1991, 176) to conclude that "the obvious publication and citation parameters for assessing research excellence do not, in this instance, seem to reproduce the clear differentiation produced by peer review." The authors go on to say that peer review of the kind associated with RAEs "tends to emphasise the comparative recognition of the leading—usually senior—members of staff," and this may, in part, explain differences in departmental rankings. A subsequent study by Oppenheim (1997) found further evidence of strong positive association.

He assessed the correlation between the scores achieved in the 1992 RAE by all departments of genetics, anatomy, and archaeology and the number of citations received by their respective faculties for the same period, based upon the ISI's citation indexes. In each case, he found statistically highly significant correlations between the RAE rank and the ranks based on both total and average faculty citation scores, leading him to conclude (1997, 477) that "citation counting provides a robust and reliable indicator of the research performance of U.K. academic departments in a variety of disciplines" and should, therefore, "be used to suggest the rank ordering, but not the actual RAE scores." Further evidence of the strong, positive correlations between RAE ratings and departmental citation scores has been furnished by Smith and Eysenck's (2002) study of U.K. psychology departments. The predictive power of citation analysis has also been explored by Holmes and Oppenheim (2001) in relation to the RAE, and more generally by Harnad, Carr, Brody, and Oppenheim (2003). Most recently, Feitelson and Yovel (2004) have used CiteSeer data to predict the citation rankings of computer scientists, focusing on a list of 10,000 top-cited authors in the field. The growing availability of large-scale digital libraries will create new citation analysis and modeling possibilities, as Kurtz et al. (2004) have shown in their revealing studies of the use made of the NASA digital library and the link between article "hits" or "reads" and subsequent citation of those articles in the open literature. We revisit this topic in chapter 8 ("The Attention Economy") and, summarily, in chapter 9.

The studies by Oppenheim, Seng, and Willett and by Zhu and Meadows, in particular, explored variations in rankings depending on whether peer review or bibliometric approaches were used. All, to differing degrees, acknowledged the utility of citation counts in determining research quality. None, however, explored the nature of the relationship between salary and high performance ratings; it is not possible to infer from their data that professors who are highly published and cited are commensurately salaried. We (Cronin and Overfelt 1994, 69) touched on the association between performance and pay in our bibliometric dissection of a U.S. information studies department and found that there was "a statistically significant correlation between faculty salaries and citation scores, though linearity should not be assumed." In the United Kingdom, however, salary data are not publicly available, as is the case for the majority of American universities, which makes comparative analysis well nigh impossible.

Citation counts can also be used (along with other pertinent data) to evaluate administrative accountability for salary variance among faculty. In a study reported in the *American Psychologist*, Gray (1983) correlated the salary levels of almost two hundred faculty at Montana State University (MSU) and the University of Montana (UM) with the following three measures: (a) years of teaching experience, (b) external recognition by way of awards for professional service or service in public affairs, and (c) citations. Correlations and multiple regressions were run for the scientists considered for the two universities separately and for each of the fifteen departments represented in the sample (humanities departments were not included in the study). He found a number of intriguing anomalies in terms of reward equity (i.e., variance that could not be accounted for by a combination of the three sets of predictors mentioned above) that led ultimately to a number of resignations or retirements at MSU, such that by the time Gray had reported his findings at a professional meeting "all administrators who could be held accountable for salary inequalities at MSU were out of office" (Gray 1983, 117).

Although there exists a large corpus of empirical studies on unexplained salary differentials between male and female faculty, productivity and merit measures have been conspicuous by their absence from much of the published research into salary equity issues (Moore 1993). Toutkoushian (1994, 62) found that "various measures of citation counts are highly correlated with faculty salaries" and that "the various citation measures reduce the unexplained salary differential between male and female faculty by 8 to 26 percent." He also noted that male faculty at the University of Minnesota are 8 percent more likely to be highly cited than female faculty when experience and other educational variables are held constant, leading him to conclude (1994, 72) that "a faculty member's gender has a significant effect on the likelihood of being highly cited." This is by no means an original or isolated finding; there is a body of empirical evidence that suggests that there is structural (gender-linked) bias in citation practices (Davenport and Snyder 1995; Ferber 1986; Lutz 1990; Moore 1993; Toutkoushian 1994).

THE SCHOLARS' MARKETPLACE

The nature of the reward system in higher education has been analyzed exhaustively by sociologists, economists, and educationalists for decades (Bourdieu 1988; Cole and Cole 1973). Few, however, would deny that the allocation of resources within academia, be they material or symbolic, is linked intimately with research performance. In a competitive market, scholars-as-authors are competing for priority and pursuing the coin of recognition. Garner (1979, 575) argues that scholarly publication "signals ability to current or prospective employers," while Feigenbaum and Levy (1995, 217) suggest that "a researcher's pursuit of professional objectives can be represented by a citation maximizing model of behaviour." In a study of the economics of fame, Levy (1988, 621) chooses to "approximate fame by citation in the modern literature" and Siow (1991, 238) demonstrates that the popular belief that first impressions are important in determining career success is theoretically sound, at least as far as career academics are concerned: "in particular early publications and citations matter the most in determining professional success of a scholar." Although all of this may seem to constitute a somewhat cynical interpretation of academic life, there is ample evidence to suggest that market signaling, linked to a publication/citation maximization strategy, whether consciously applied or not, can translate into significant economic gains for the successful perpetrator. That, of course, is not to say that early publications and citations may not simply indicate extraordinary talent.

The maximization strategy, in fact, finds popular and partial expression in the admonition "Publish or perish," the truth of which, whilst it may vary to some extent depending on the nature of the university in question, is nonetheless empirically verifiable. As Boyes, Happel, and Hogan (1984, 140) found in their survey of North American economics departments, "research and teaching are weighted differently (in promotion/tenure decisions) depending on the market segment in which an institution operates." Broadly similar findings emerged from Kasten's (1984) study of the factors that affected tenure and merit pay decisions at the University of Nebraska. The importance of publication—more specifically, the number of articles accepted by top-ranking journals— has been demonstrated conclusively in a range of studies seeking to measure relative rates of return for a battery of performance indicators.

In their study of the reward structure in the University of Wisconsin's economics department, Siegfried and White (1973, 316) found that the "statistical significance and quantitatively larger impact of research output appears to support the popular hypothesis that the faculty reward structure at large public universities encourages research, possibly at the expense of teaching." Specifically, they were able to assign different monetary values to different categories of publication. Monographs, for instance, have no statistically significant effect on salary levels, whereas each article in one of "the six national, general journals is 'worth' an additional $392 a year" (1973, 313).

Confirmation of these results was provided by Sauer (1988, 864), whose survey of one hundred and forty academic economists shows that "there are significant monetary returns to high-quality research in the economics profession." His approach was taken a step further by Tuckman and Leahey (1975, 96) in their self-referential paper, "What is an article worth?" They developed a method to compute lifetime returns for selected article categories by academic rank, and applied it to a national sample of almost one thousand economists in North America. In a novel extension of this approach, Laband and Tollison (2000, 649) computed "the dollar-equivalent value-added by colleagues' comments" and further found that the value of receiving informal comments and feedback was, in part, a function of the trusted assessor's citation stock. Their conclusion (2000, 651) confirms what we know intuitively to be the case: "Little wonder then that young economists often hitch their wagons to senior people in a mentoring process."

PECUNIARY CORRELATIVES OF CITATION

Investigating the determinants of income among more than three thousand recipients of natural science doctorates between 1957 and 1959, Holtmann and Bayer (1970, 412) included "citation count" as a variable in their regression analysis, arguing that the "number of times a person's work is cited is probably the best single indicator of his scientific productivity." They found (1970, 416) that a researcher's citation count "shows a significant positive correlation with income." Hamermesh, Johnson, and Weisbrod (1982, 473) took as their starting point the proposition that academe's uniqueness (in respect of pay determination) is that it "consists of a community of scholars whose physical locations may be far apart, but who participate together in the production of knowledge." As a consequence, "one scholar's social productivity should be measured by the sum of direct and indirect influences on other producers as well as by direct contributions (publications)." The number of references to a scholar's work thus functions as a proxy for influence. In this social-welfare variant of the maximizer strategy mentioned earlier, scholars who attract a large number of citations should be rewarded preferentially by the market. They discovered (1982, 481) that an additional citation "adds more to salary than does the publication of an additional book or article."

The notional amount, of course, will vary depending upon career point, an individual's net current citation stock, and the discipline or subject area in question. Nonetheless, Diamond (1986a, 200, 202) was able to specify the pecuniary value of citations: "The marginal value of a citation (when the level of citation is zero) varies between $50 and $1,300," though it is to be expected that "the marginal value of an additional citation diminishes as the level of citation increases." His conclusions are based partly on the results of a study of mathematicians at the University of California, Berkeley (Diamond 1985), in which he found that citations "to multiple-authored papers are worth more than citations to singly authored work no matter what the order of the author's name." This finding has been confirmed with a sample of economics professors by Sauer (1988). How might this counterintuitive result be explained? Diamond (1985, 319) speculates that "citations to multiple-authored articles are a proxy for the trait of collegiality that is rewarded by departments in the determination of salaries." Rousseau (1992, 79), on the other hand, has offered a simple Bayesian proof "without recurrence to 'quality' arguments" to explain why multiauthored papers are more cited than others.

From a methodological standpoint, Diamond (1986a, 212) provides reinforcement for those who maintain that first-author searches run against ISI's citation indexes provide sufficiently robust data to make costly and time consuming coauthor searching unnecessary: "The coefficient of determination for a regression including a measure of citations to nonfirst-author math articles as a regressor was the same as the coefficient of determination for a regression that omitted the regressor." This conclusion, in fact, provides support for Oppenheim's (1995,

20–21) intuitive feeling that no significant distortions were introduced into his rankings as a result of restricting the citation analysis to first authors only. Given the relative ease with which first-author citation searching can be conducted, the potential efficiency gains should appeal to bodies such as the HEFC in the United Kingdom.

CONCLUSIONS

The evidence from a range of studies conducted in North American universities on data relating to scholars in (principally) scientific disciplines, or subjects at the hard end of the social sciences spectrum, suggests that citations function as reliable predictors of pecuniary success within the academic reward system, this despite persistent concerns about what a citation signifies (Simkin and Roychowdhury 2003). As Diamond (1986a, 212; and see also Diamond 1986b, 2000) notes, "A robust finding in all of the studies . . . is that citations are indeed a positive and significant determinant of earnings over almost all of the observed range of citation levels." Research in the United Kingdom confirms the potential utility of citations as indicators of scholarly productivity. The weight of transatlantic evidence suggests that citation counts have a role to play in the context of national, institutional, and programmatic research assessment exercises. Given the strength of the observed associations between citation counts on the one hand and both scholarly impact and career success on the other, it is somewhat surprising that more use is not routinely made of citation data by both universities and funding agencies in the evaluation of research productivity and effectiveness.

REFERENCES

Alonso, C. J. (2003). Editor's column: having a spine—facing the crisis in scholarly publishing. *Proceedings of the Modern Language Association*, 118(2), 217–223.

Andersen, D. L. (Ed.). (2003). *Digital Scholarship in the Tenure, Promotion, and Review Process*. Armonk, NY: Sharpe.

Anderson, A. (1991). No citation analyses please, we're British. *Science*, 252, 639.

Bourdieu, P. (1988). *Homo Academicus*. Cambridge: Polity Press.

Boyes, W. J., Happel, S. K., and Hogan, T. D. (1984). Publish or perish: Fact or fiction? *Journal of Economic Education*, 15(2), 136–141.

Burnhill, P. M. and Tubby-Hille, M. E. (1994). On measuring the relation between social science research activity and research publication. *Research Evaluation*, 4(3), 130–152.

Burt, R. S. (1992). *Structural Holes: The Social Structure of Competition*. Cambridge, MA: Harvard University Press.

Cole J. and Cole. S. (1973). *Social Stratification in Science*. Chicago: Chicago University Press.

Collini, S. (1989). Publish—and be dimmed. *Times Higher Education Supplement*, March 2, 13–14.

Cronin, B. and La Barre, K. (2004). Mickey Mouse and Milton: Book publishing in the humanities. *Learned Publishing*, 17(2), 85–98.

Cronin, B. and Overfelt, K. (1994). Citation-based auditing of academic performance. *Journal of the American Society for Information Science*, 45(2), 61–72.

Cronin, B. and Overfelt, K. (1995). E-journals and tenure. *Journal of the American Society for Information Science*, 46(9), 700–703.

Davenport, E. and Snyder, H. W. (1995). Who cites women? Who do women cite?: An exploration of gender and scholarly citation in sociology. *Journal of Documentation*, 51(4), 404–410.

David, P. A. and Spence, M. (2003). *Towards Institutional Infrastructures for E-Science: The Scope of the Challenge*. Oxford: Oxford Internet Institute.

Davidson, C. N. (2003, October 3). Understanding the economic burden of scholarly publishing. *Chronicle of Higher Education,* B7–10.

Diamond, A. M. (1985). The money value of citations to single-authored and multiple-authored articles. *Scientometrics*, 8(5–6), 315–320.

Diamond, A. M. (1986a). What is a citation worth? *Journal of Human Resources*, 21, 200–215.

Diamond, A. M. (1986b). The life-cycle research productivity of mathematicians and scientists. *Journal of Gerontology*, 41(4), 520–525.

Diamond, A. M. (2000). The complementarity of scientometrics and economics. In: Cronin, B. and Atkins, H. B. (Eds.). *The Web of Knowledge: A Festschrift in Honor of Eugene Garfield*. Medford, NJ: Learned Information, 321–336.

Estabrook, L. (2003). *The Book as the Gold Standard for Tenure in Humanistic Disciplines*. Available at http://lrc.lis.uiuc.edu/reports/cic/CICBook.html.

Feigenbaum, S. and Levy, D. M. (1995). The market for (ir)reproducible econometrics. *Social Epistemology*, 7(3), 215–232.

Feitelson, D. G. and Yovel, U. (2004). Predictive ranking of computer scientists using CiteSeer data. *Journal of Documentation*, 66(1), 44–61.

Ferber, M. (1986). Citations: Are they an objective measure of scholarly merit? *Signs*, 11, 381–389.

Garfield, E. (1979). *Citation Indexing: Its Theory and Application in Sciences, Technology, and Humanities*. New York: Wiley.

Garfield, E. (1988, October 31). Can researchers bank on citation analysis? *Current Contents*, 44, 3–12.

Garner, C. A. (1979). Academic publication, market signaling, and scientific research decisions. *Economic Inquiry*, 17, 575–584.

Gray, P. H. (1983). Using science citation analysis to evaluate administrative accountability for salary variance. *American Psychologist*, 38, 116–117.

Greenblatt, S. (2002). Call for action on problems in scholarly book publishing; a special letter from Stephen Greenblatt. Available at http://chronicle.com/jobs/2002/07/2002070202c.htm.

Hamermesh, D. S., Johnson, G. E., and Weisbrod, B. A. (1982). Scholarship, citations and salaries: Economic rewards in economics. *Southern Economics Journal*, 49(2), 472–481.

Harnad, S., Carr, L., Brody, T., and Oppenheim, C. (2003). Mandated online RAE CVs linked to university eprint archives: Enhancing U.K. research impact and assessment. *Ariadne*, 35. Available at www.ariadne.ac.uk/issue35/harnad/intro.htm.

Hodges, S., Hodges. B., Meadows, J., Hancock-Beaulieu, M., and Law, D. (1996). The use of an algorithmic approach for the assessment of research quality. *Scientometrics*, 35, 3–13.

Holmes, A. and Oppenheim, C. (2001). Use of citation analysis to predict the outcome of the 2000 research assessment exercise for unit of assessment (UoA) 61: library and information management. *Information Research*, 6(2). Available at http://informationr.net/ir/6-2/paper103.html.

Holtmann, A. G. and Bayer, A. E. (1970). Determinants of professional income among recent recipients of natural science doctorates. *Journal of Business*, 43, 410–418.

Jaffe, S. (2002, November 11). Citing UK science quality. *Scientist*, 16(52). Available at www.the-scientist.com/yr2002/nov/prof1_021111.html.

Jeschin, C., Dawson, G., and Anderson, J. (1995). A bibliometric database for tracking acknowledgements of research funding. In: Koenig, M. E. D. and Bookstein, A. (Eds.). *Proceedings of the Fifth Biennial Conference of the Society for Scientometrics and Infometrics*. Medford, NJ: Information Today, 235–244.

Kasten, K. L. (1984). Tenure and merit pay as rewards for research, teaching, and service at a research university. *Journal of Higher Education*, 55(4), 500–514.

Kennedy, D. (1994). Making choices in a research university. In: Cole, J. R., Barber, E. G., and Graubard, S. R. (Eds.). *The Research University in a Time of Discontent*. Baltimore: Johns Hopkins University, 85–115.

136 *Chapter Six*

Kling, R. and Spector, L. B. (2003). Rewards for scholarly communication. In: Andersen, D. L. (Ed.). (2003). *Digital Scholarship in the Tenure, Promotion, and Review Process*. Armonk, NY: Sharpe, 78–103.

Kurtz, M. J., Eichorn, G., Accomazzi, A., Grant, C., Demleitner, M., Murray, S. S., Martimbeau, N., and Elwell, B. (2004). The bibliometric properties of article readership information. *Journal of the American Society for Information Science and Technology* (in press).

Laband, D. N. and Tollison, R. D. (2000). Intellectual collaboration. *Journal of Political Economy*, 108(3), 632–662.

Leatherman, C. (1995, February 3). Credentials on trial. *Chronicle of Higher Education*, A14, A16.

Levy, D. M. (1988). The market for fame and fortune. *History of Political Economy*, 20(4), 615–625.

Lewis, P. (2002). Is monograph tyranny the problem? *Proceedings of the Modern Language Association*, 117, 1222–1224.

Lohmann, S. (2002). Herding cats, moving cemeteries, and hauling academic trunks: Why change comes hard to the university. Paper presented at the *Annual Meeting of the Association of the Study of Higher Education*, Sacramento, California, November 21–24, 2002.

Lutz, C. (1990). The erasure of women's writing in sociocultural anthropology. *American Ethnologist*, 17, 611–627.

March, J. G., Schulz, M., and Zhou, X. (2000). *The Dynamics of Rules: Change in Written Organizational Codes*. Stanford, CA: Stanford University Press.

Martin, B. R. and Irvine, J. (1983). Assessing basic research: Some partial indicators of scientific progress in radio astronomy. *Research Policy*, 12, 61–90.

Monaghan, P. (2004). Presses seek fiscal relief in subsidies for authors. *Chronicle of Higher Education*, 50(49), A1, A14–15.

Moore, N. (1993, February). Faculty salary equity: Issues in regression model selection. *Research in Higher Education*, 34, 107–126.

Oppenheim, C. (1995). The correlation between citation counts and the 1992 research assessment exercise ratings for British library and information science university departments. *Journal of Documentation*, 51(1), 18–27.

Oppenheim, C. (1997). The correlation between citation counts and the 1992 research assessment exercise ratings for British research in genetics, anatomy and archaeology. *Journal of Documentation*, 53(5), 477 487.

Regier, W. G. (2003, June 13). 5 problems and 9 solutions for university presses. *Chronicle of Higher Education*. Available at http://chronicle.com/prm/weekly/v49/i40/40b00701.htm.

Readings, B. (1996). *The University in Ruins*. Cambridge, MA: Harvard University Press.

Rousseau, R. (1992). Why am I not cited or why are multi-authored papers more cited than others? *Journal of Documentation*, 48, 79–80.

Ruark, J. K. and Montell, G. (2002, July 2). A wake-up call for junior professors. *Chronicle of Higher Education.* Available at http://chronicle.com/jobs/2002/07/2002070201c.htm.

Ryan, J. et al. (2002). The future of scholarly publishing. *Profession 2002.* New York: MLA, 172–186.

Sauer, R. D. (1988). Estimates of the returns to quality and coauthorship in economic academia. *Journal of Political Economy*, 96(4), 855–866.

Scholes, R. (2004, Summer). The evaluation of faculty members in the culture of "excellence." *MLA Newsletter*, 3.

Seng, L. B. and Willett, P. (1995). The citedness of publications by United Kingdom library schools. *Journal of Information Science*, 21(1), 68–71.

Siegfried, J. J. and White. K. J. (1973). Financial rewards to research and teaching: a case study of academic economists. *American Economic Review*, 63, 309–316.

Simkin, M. V. and Roychowdhury, V. P. (2003). Read before you cite! *Complex Systems*, 14, 269–274.

Siow, A. (1991). Are first impressions important in academia? *Journal of Human Resources*, 26(2), 236–255.

Smith, A. and Eysenck, M. (2002, July). The correlation between RAE ratings and citation counts in psychology. Available at http://psyserver.pc.rhbnc.ac.uk/citations.pdf.

Swanson, D. R. (1989). A second example of mutually isolated medical literatures related by implicit, unnoticed connections. *Journal of the American Society for Information Science*, 40(6), 432–435.

Toutkoushian, R. K. (1994). Using citations to measure sex discrimination in faculty salaries. *Review of Higher Education*, 18(1), 61–82.

Tuckman, H. P. and Leahey, J. (1975). What is an article worth? *Journal of Political Economy*, 83, 951–968.

Warner, J. (2000, October 30). Research assessment and citation analysis. *Scientist*, 14(21), 39.

Waters, L. (2001, April 20). Rescue tenure from the tyranny of the monograph. *Chronicle of Higher Education* Available at http://chronicle.com/free/v47/i32/32b00701.htm.

Waters, L. (2004). *Enemies of Promise: Publishing, Perishing, and the Eclipse of Scholarship*. Chicago: Prickly Paradigm Press.

Zhu, J. and Meadows, A. J. (1991). Citations and departmental ratings. *Scientometrics*, 21(2), 171–179.

Chapter Seven

Symbolic Capitalism

The accumulation of symbolic capital is a driving force of academic life. Successful capital formation is commonly signified by the trappings of scholarly distinction (e.g., membership of national academies, receipt of honorary doctorates) or acknowledged status as a public intellectual. Validity and reliability concerns notwithstanding, citations are widely used as indicators of peer esteem or proxies for intellectual influence and impact (Cronin 1984)—and also viewed as expressions of social trust (Davenport and Cronin 2000). And as we saw in the previous chapter, the rates of return to citation are substantial in the context of the academic reward system. Bourdieu (1990, 76) has referred to citations as the "most objectified of the indices of symbolic capital," while Nelson (1997, 39) has likened them to "academia's version of applause." Elsewhere, I (Cronin 2000, 450) have referred to the "transvaluation of these stockpiles of manipulated and manipulable capital into objectified ratings and rankings." Within the political economy of research universities, citations are a highly regarded form of symbolic capital; in the vernacular, you can bank on citations, though analogies to the market tend to overlook the importance of status hierarchies, social closure, and social capital in explaining "the structure and dynamics of academic employment" (Burris 2004, 244). However, ratings and indices of perceived intellectual worth are not unproblematic, as Bourdieu (1986, 1990, 1991) and others (Sosteric 2000) have noted; nor should we underestimate the complexity of discursive practice—of which citation is an important facet—within scholarly communities (Budd 2001; Hyland 2000).

With the advent of the web, we now have new ways of tracking scholars' visibility, both within and beyond their traditional spheres of influence. In an early discussion of web-derived indicators of scholarly salience, we (Cronin et al. 1998, 1326) made the following point: "While traditional citation analysis can tell us a lot about the formal bases of intellectual influence, it, quite naturally, tells us nothing about the many other modalities of influence which comprise the total impact of an individual's ideas, thinking and general professional presence." We went on to say (1998, 1326), "[t]he range of genres of invocation made possible by the Web should help give substance to modes of influence which have historically been backgrounded in narratives of science." To demonstrate our point, we developed an inductive typology of web-based invocations to cover the spectrum of mentions (from a simple listing of one's name as a faculty member, through inclusion of one's textbook on a course syllabus, to being the center of a serious scholarly discussion on a listserv) based upon the form or context in which the mentions occurred. The eleven categories we used were abstract, article, conference proceedings, current awareness, external home page, listserv, personal/parent organization home page, resource guide, book review, syllabus, and table of contents. Of course, receiving a mention—or making a splash—on the web is not quite the same thing as being cited by one's academic peers in a refereed journal or scholarly monograph; the diversity of the web audience, along with the broader repertoire of possible forms of invocation and associated motivations, calls for interpretive caution. Specifically, the population of potential "citers" on the web is broader than one's habitual, or even extended, scholarly audience. However, the web provides a means for those beyond the traditional discursive pale to both receive and react to ideas once corralled within the formal scholarly literature. It also provides academics with the means of reaching a wider audience for both their research and courseware, well beyond their immediate institutions and disciplines. Thus, one needs at the very least to distinguish between, on the one hand, enduring scholarly impact—as suggested by a cumulating citation record—and, on the other, web-based measures of "transient group interest . . . a digital-age equivalent of Andy Warhol's fifteen minutes of fame for web-present authors" (Cronin 1999, 954). That said, it would be perverse to preclude web-derived measures of impact, celebrity, or esteem from consideration in academic evaluation exercises. Some emerging indicators of online recognition and recommendation may, in time, prove to be very bankable forms of symbolic capital.

Posner (2001, 167) has argued that web hits and media mentions can be used conjointly as proxies for "public-intellectual status." Building on an earlier study of legal scholars as public intellectuals (Landes and Posner 2000), he undertook a comparison "of a public intellectual's academic renown, as proxied by the number of scholarly writings, to his celebrity as a public intellectual, as proxied by media mentions" (2001, 169). For the first measure he used citation scores based on ISI's citation indexes; for the latter he relied on both web hits and mentions in LexisNexis general news sources. Posner used a large, heterogeneous sample (N=546) drawn from all walks of academic and public intellectual life (e.g., Hannah Arendt, Allan Bloom, William F. Buckley, Umberto Eco, Milton Friedman, Stephen Jay Gould, Conor Cruise O'Brien, Susan Sontag, Lionel Trilling, Edward Wilson). Moreover, he meticulously documented his data collection and analysis methods (2001, 188–193) and made the entire dataset freely available on the web to interested researchers.

The world of citation is the closed world of the clerisy; we trade citations with other scholars, not with the public at large. The world of the web, by contrast, is more open and egalitarian in character (equal opportunity invocation, if you will); here, we are mentioned/linked to by our peers but also, on occasion, by professionals and practitioners and, indeed, sundry others who may have a special or passing interest in some of the issues we address in our lives as academics and/or public intellectuals. The web extends the discursive space within which scholars operate. What these polyvalent invocations mean in aggregate, or individually, is another matter, and it is doubtful if, unprocessed and unwinnowed, they should be included among the battery of partial indicators conventionally used to evaluate scholarly production and impact. The media world is anything but closed (hence, presumably, the very positive correlation between the web and media indicators); it is here, if we are lucky, that we amass celebrity capital through our strategically placed sound bites and on-tap expertise.

We (Cronin and Shaw 2002) replicated Posner's study with a sample of information studies (IS) professors. Our data—the aggregate media mentions—seem to suggest that there are no outstanding public intellectuals in the IS field; whether that is a function of our collective diffidence or an expression of the media's/public's estimation of our worth is an open question, one requiring comparative analyses of different disciplinary and professional groups. What we can say with some assurance is that the supposedly leading scholars in IS are unseen and unheard in the public sphere; unlike some other disciplines (e.g., law, whose public faces include notables such as Dershowitz, MacKinnon, Posner, Strossen, and Tribe), the IS field has yet to bring forth, to use the prevailing terminology, its first bona fide "academostar" (Spurgin 2001) or public intellectual.

SEMIOSIS

I now return to citations and offer a semiotic analysis of these signaling devices in academic writing. By way of a preface, I should note that in this chapter I make a subtle ontological distinction between references and citations. Elsewhere in this book I have and shall continue to use "reference and "citation" and also "referencing" and "citing" interchangeably, as, indeed, is common practice in both professional discourse and scientific writing. What follows has eight component parts: (1) the reciprocal relationship between bibliographic references and citations; (2) triadic sign systems; (3) dominant theoretical frameworks; (4) bibliometrics and psychological relevance; (5) the communicative properties of references; (6) bibliographic references as concept markers or symbols; (7) referencing behavior and reader response; (8) symbolic capital and cyber-surveillance.

I begin anecdotally. Sebeok's (1994) essay *The Study of Signs* opens with ten little dramas wherein each of the subjects, be it a radiologist or hunter, goes about his business, diagnosing a silhouette on a chest X-ray, deciphering animal tracks in the snow, whatever is occupationally or avocationally apposite. For the semiotician, these vignettes all have something in common, what Sebeok (1994, 4) calls "sign action." Identifying, classifying, and interpreting sign actions are the constitutive tasks of semiotics—the science of signs. These tasks also lie at the heart of citation analysis, where a sign is not just a sign. Reading Sebeok reminded me of a phrase I had used many years ago in a review article on citation theory (Cronin 1981): "Citations are frozen footprints in the landscape of scholarly achievement; footprints which bear witness to the passage of ideas." Just as with Sebeok's hunter who surmises from the tracks in the snow that a herd of elk is moving ahead of him, so from citational "footprints it is possible to deduce direction; from the configuration and depth of the imprint it should be possible to construct a picture of those who have passed by, whilst the distribution and variety furnish clues as to whether the advance was orderly and purposive." In retrospect it is odd that a paper entitled "The Need for a Theory of Citing" should make no mention of semiotics. I also failed to make an explicit distinction between references and citations, though for many members of the loosely defined bibliometrics/scientometrics community this remains a seemingly esoteric and unnecessary distinction, a "merely technical difference hardly relevant for anyone but he who is inherently meticulous" (Wouters 1998, 231). Here, I want to demonstrate the importance of distinguishing clearly between these two constructs. To this end, I endeavor to outline a conceptual framework capable of accommodating the "constitutional complexity" of citation (Leydesdorff 1998, 9).

POLYSEMY

Within information science, there is voluminous research literature, theoretical and applied, devoted to citation analysis and referencing practice, but with few exceptions (Brier 2004) this corpus does not acknowledge or interact with the core literature of semiotics. These two academic tribes seem ignorant of one another's existence, though there is some evidence of cross-fertilization and the gradual formation of a common discourse community (Gluck 1997; Wouters 1998). But what are references if not instances of sign action—a way of pointing, ultimately and unambiguously, to specific texts in the public domain? It is a two-step process: the embedded reference is a pointer to the full bibliographic record at the end of the paper, which itself is a pointer to the text in question. At face value, a reference is a signaling device, a way of communicating to the reader that I am familiar with, and have drawn upon, a particular work of a particular author. However, referencing may also be interpreted as a strategizing device (Ben-Ari 1987), a means of locating one's thesis in a particular intellectual milieu, or, less charitably, as a weakly masked attempt to imply some degree of social

familiarity, celebrity endorsement, or coat-tailing—part of what Bourdieu, Passeron, and Saint Martin (1984, 20) call the "propitiatory ritual
of erudite citation [that] pays homage to celebrated masters or to culture." Multiple interpretations of referencing behaviors and their extratextual import are thus possible.

Acknowledgments likewise testify to cognitive influence but, additionally, they may bear witness to technical, procedural, moral, and financial support proffered by myriad individuals and institutions. There
are important differences, though, which need to be recognized at the
outset (Cronin 1995, 21):

> Both citations and acknowledgements declare a relationship . . . which
> may be profound or superficial. One, the citation, has objective status in
> that a third party can refer to the cited document and corroborate the cit-
> ing author's interpretation, pursue an intellectual lead, or chain backwards
> or forwards through the related literature; the other, the personal ac-
> knowledgement, describes an inherently private interaction, or debt,
> which, by definition, cannot have the same commodity status.

Prima facie, there are several instrumental and rhetorical similarities between references and acknowledgments, such that if either one is to be
subjected to semiotic scrutiny, then both should be exposed to the same
degree of forensic examination. References and acknowledgments,
along with citations, are first cousins in an extended family of signs, and
the natural focus of the bibliometrics research community.

In passing, it is worth noting that this set of established signs (references, acknowledgments, citations) may have to accommodate additional genres of signaling behaviors reflecting scholars growing use
of the World Wide Web. The web is giving rise to new modes of communication. The ways in, and reasons for, which individual researchers and scholars are mentioned on (linked to) the web are mul-
tifaceted. It is conceivable that novel forms of citation (generically
speaking) will evolve, which could be used as indicators of cognitive
or social influence within specific disciplines or communities of professional practice (Rousseau 1997). The potential significance of
these different genres of invocation is a subject deserving of investigation and constitutes a logical progression of research into the epistemological and normative bases of citation behavior, as we shall see
in the next chapter.

From a semiotic perspective, the patterning of references throughout this monograph proclaims a web of connections—"transtextual relationships" (Genette 1997, 1)—between the cited and citing authors' works; they reveal "a trace of conversations between texts" (Czarniawska-Joerges 1998, 63). Such invocation is typically assumed to imply some measure of cognitive correspondence, ideational interplay, between the citing and cited texts. What that is, how it is apprehended and used by others (i.e., reader response), and whether the assumption itself is valid remain, of course, contestable issues. And herein lie the fundamental problems associated with both referencing practices and citation counting—slippery semantics and invisible intention.

REFERENCES AND CITATIONS: A RECIPROCAL RELATIONSHIP

Perhaps the best way of bringing to light the difference, or the reciprocal relationship, between references and citations is to quote verbatim from a thoughtful article by Wouters (1993, 7) that seeks to connect theory lines from semiotics and information science. Subtle though the distinction may appear, it has significant entailments in the context of both faculty and programmatic assessment exercises:

> In summary, the sign <reference> is a pointer belonging to the citing text. It points to the cited text but is still an attribute of the citing text. The sign <citation>, however, constructed by inverting the reference, is an attribute of the cited text. The cited text is the (absent) referent of the scientometric citation. Thus the two have different referents and can consequently best be analyzed as two different signs.

Elsewhere, Wouters (1998, 233) expands on the relationship between the two:

> If reference R of citing article A points at article B, the corresponding citation C is initially nothing else than a different format of reference R. The citation is the mirror image of the reference. This rather innocently looking inversion has important consequences. By creating a different typographical format of the lists of references—by organizing the references not according to the texts they belong to, but according to the texts they point at—they become attributes of the cited instead of the original, citing texts.

What do references mean when they are inverted and redesignated as citations? They are variously interpreted as records of intellectual trading, or as evidence of peer interactive communication (McCain 1991), which testify to instrumental or other kinds of influence. Converted into citations, they are widely held to be measures of intellectual impact, indicators of perceived utility or surrogate measures of academic quality. Are they some or all of these things, in some or all contexts, at some or all times? In reviewing the literature (e.g., Liu 1993), one is struck by the degree to which a variety of social practices and associated explanations have been confounded and also by the extent to which mutual incomprehensibility and distrust seem to have characterized the debate between proponents and opponents of citation analysis (e.g., Edge 1977; White 1990).

There is a strong preference for metaphorical explanations (e.g., "scholarly bricklaying" [Price 1963, 64]) and many writers exhibit an inability to disambiguate references, acknowledgments, and citations, as Egghe and Rousseau (1990) have shown. How do we deal with the fact that references can have multiple articulations, that they function as signs that afford "interpretative flexibility" (Wouters 1993, 7)? What does it mean to be heavily cited and, conversely, what does it say about a scholar whose work lies uncited? What cultural meaning is ascribed to citedness, and does this vary across epistemic cultures and communities of practice? These are not trivial questions, and the problem is magnified when aggregations of citations are analyzed and used as the basis of institutional or individual evaluation, a now well-established, though by no means universally welcomed, phenomenon in higher education.

Although the figures vary from discipline to discipline, it is generally true that many published papers are not cited even once five years after their date of publication (Garfield 1998). In a sense, citation analysis privileges a minority of published journal articles. Every paper is a reference waiting to happen; once activated (i.e., referenced), that paper acquires a polyvalent semiotic character. It is both the object pointed to by the textual reference and an indicator of presumptive influence; it is both sign and symbol. These indicators of influence are selectively marshaled in the form of the *Science Citation Index* (and its sister products) to create a cumulating ledger of scholars' symbolic market worth. The relatively recent epiphenomenon of citation has generated an "evaluation industry" (Trow 1998, 25) of its own. In the next section, I show how this communicative convention has been subjected to progressive commodification.

TRIADIC SIGN SYSTEMS

References and citations are not simply different signs, the point made by Wouters, but different sign systems. Gluck (1997, 53) has provided an overview of how semiotic analysis could be used to inform our understanding of information behaviors, employing Peirce's sign triad (sign vehicle, interpretant, referent) as his root typology (see figure 7.1). Although alternative formulations and terminology are possible, as he notes, this triad allows us to examine references and citations in terms of three common dimensions: (a) the carrier of meaning (sign-vehicle), (b) the meaning or concept referred to (interpretant), and (c) the object pointed to (referent). How, then, do these elements reveal themselves with each class of sign (reference, acknowledgment, or citation)? Figures 7.2 through 7.5 demonstrate the distinctiveness of the various sign systems at work, and underscore the need for requisite interpretative variety and expositional clarity.

In figure 7.2, the embedded reference is the sign-vehicle. It has dual referents: the full bibliographic reference at the end of the paper and the object for which that reference is a surrogate, the cited work. The interpretant is the meaning or concept flagged by the sign-vehicle. It may be clearly grounded (for example, a formula is quoted) or ambiguous (a

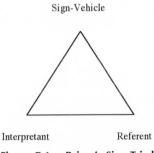

Figure 7.1. Peirce's Sign Triad (after Gluck, 1997)

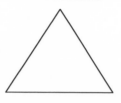

Figure 7.2. **Bibliographic Reference Sign Triad**

global reference is provided to an individual's oeuvre), yet reader response theory is largely absent from critical analyses of citation behavior: I return to the "politics of interpretation" (Wouters 1998, 226) later.

In figure 7.3, we can see that the referent of an acknowledgment statement is an interaction or event of some kind that involved the author and another social actor. This referent lacks the objectified (verifiable) status of the referent in figure 7.2, but the debt it records (the interpretant in this instance) may be as important or even more so than that implied by a bibliographic reference.

The picture is different with respect to the citation. In this case (figure 7.4), the sign-vehicle is typically found in the *Science* (or *Social Sciences*) *Citation Index* (*SCI/SSCI*) and detached from its referent, the paper that it denotes, and the related content. As Wouters (1998, 232) observes, the "basic function of the Science Citation Index (and similar

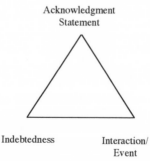

Figure 7.3. **Acknowledgment Sign Triad**

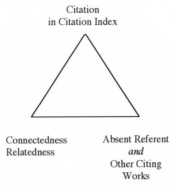

Figure 7.4. **Citation Sign Triad**

devices) is to turn an enormous amount of lists of references upside down." The citation in fact has multiple referents, namely, all the papers (objects) that the *SCI/SSCI* list as having invoked it. The citation points back to its parent article and forward to the population of papers that have referenced it over time. Its referents are multiply articulated. The meaning of these signs is best understood in terms of the intertextual relationships posited between the citing and cited papers and in the social networks and maps of science that they make manifest.

When aggregations of citations are analyzed, the picture changes yet again (see figure 7.5). In this case, the interpretant is typically expressed in terms of impact, worth, or esteem, while the referent becomes the author. In evaluative bibliometrics, the focus is much more likely to be

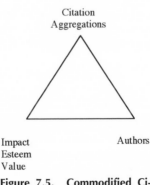

Figure 7.5. **Commodified Citation Sign Triad**

highly cited authors than their published works. As far as acknowledgments are concerned, there is at present no commercial equivalent of the
Science Citation Index, though the idea has been suggested and an early
design blueprint proposed (Cronin and Weaver-Wozniak 1993). Given
the vagaries of referencing practice and the range of interpretative
repertories that exists, what prospect is there of developing a unified
theory of referencing/citation? Broadly speaking, three theoretical
frameworks dominate current discussion: the functionalist, phenomenological, and normative. Both semiotics and reader-response theory are
absent as explicit framing devices.

The functionalist approach stresses the instrumental or communicative roles that references play within the text—purposive signing. Numerous functions have been identified (e.g., furnishing historical context, providing supporting evidence) and a variety of typologies and
role operators has been developed to accommodate different classes of
references (see Cronin [1984] for an overview of schemata). However,
even if we agree on the manifest purpose of a reference in context, there
remains the question as to why that particular work (with its particular
signification, what is termed *situated meaning* in figure 7.2) rather than
another was selected. The fact that referencing practices are molded by
social-psychological variables makes the matter of citer motivation
vexingly difficult to establish with certitude. What does a sign mean?
How are references imbued with meaning, and how is the intended
meaning extracted by the reader? The attraction of the phenomenological approach rests on the residual subjectivity of referencing behavior.
Nonetheless, it is clear that authors' practices, in fact, display a striking
degree of consistency. This is the basis of the normative interpretation.

Referencing behavior is governed by a more or less codified body of
rules that determine the process and protocols for sign placement; it is
anything but arbitrary. Rather, it is bound up "with the ideology and
technical practices" (Grafton 1997, 5) of the academic profession and
linked inextricably to the scientific reward system. Referencing and acknowledging are universally recognized means of dispensing credits
and repaying intellectual debts within the primary communication
process, part of what might be termed the "apparatus of witness" (Rotman 1987, 8). As far as referencing is concerned, and at the risk of stating the obvious, the reader trusts that the author has read or consulted
the cited work and that the selection of that work is grounded in seri-

ous scholastic reasons, rather than driven by personal or institutional politicking. In that sense, the business of referencing may be viewed as a series of transactions (token exchange) that can be analyzed within the context of a political and moral economy of citation (Cronin 1998a). Think, then, of references as signs of mutual trust operationalized as expressions of indebtedness (see figure 7.3). Or, as Fish (1989, 164) puts it,

> The convention is a way of acknowledging that we are engaged in a community activity in which the value of one's work is directly related to the work that has been done by others; that is, in this profession you earn the right to say something because it has not been said by anyone else, or because while it has been said, its implications have not yet been spelled out.

Too often, the functionalist, phenomenological, and normative perspectives (about which I'll say more in the next chapter) are positioned as being mutually antagonistic. This is unnecessary and unhelpful. The sign triads, on the other hand, reflect the complexity of referencing and citation behaviors, and this should suggest that shoehorning all of these practices into a single interpretative framework is doomed to failure. In fact, one of the attractions of citation analysis is that it allows for movement between the cognitive, textual, and social dimensions of science (Leydesdorff 1998). Given the seeming futility and overt hostility of much of the current debate, it is hardly surprising that naïve rationalism continues to exercise considerable appeal. White (1990, 91) has little patience with extreme relativism, preferring to believe that aggregations of these signs can tell us something about the trafficking of influence within and across disciplinary communities: "What the anticitationists generally miss is the liberating effect of using data from very large files, such as those available from ISI. When one sees that scores, hundreds, and even thousands of citations have accrued to a work, an author, a set of authors, it is difficult to believe that all of them are suspect."

RELEVANCE REVISITED

The relationship between sign and object can be difficult to pin down with precision; sometimes the bibliographic reference may be pointing to an author's oeuvre (the writings of Freud), sometimes a particular opus (Merton's review article), sometimes a motif or theme (redemption in Wagner), sometimes a particular segment of a paper (the methods section), or sometimes a quantum (a proof, motif). Intuitively, there are various levels or gradations of referencing, ranging from the subatomic through molecular to compound, as I have suggested elsewhere (Cronin 1994). However, textual analysis may not lay bare the exact relationship between the cited and citing texts. MacRoberts and MacRoberts (1986, 156) have demonstrated that many citations simply do not register with the cited work. Registration, in fact, can range from weak to strong, which leads them to conclude that citation analysis is prone to systematic bias. Elsewhere, they (MacRoberts and MacRoberts 1989) identified and classified discrepancies between influence as evident in a set of fifteen papers dealing with the history of genetics and influence as captured in the bibliographies attached to those papers. In their expert judgment, authors were seriously underreferencing sources of influence; on average, the authors whose works they scrutinized achieved only 30 percent coverage of material influences. To put it another way, when it comes to sign selection, authors get it wrong more often than not. In a letter replying to Garfield's (1997) criticisms of their position, MacRoberts (1997, 963) reiterated his reasons for opposing the use of citations as data for evaluative purposes. Revealingly, he was coauthor of six of the nine supporting references, but elected not to demonstrate that the references he had affixed to the letter provided comprehensive, pertinent, and balanced coverage of the points in his argument.

Referencing is a complex phenomenon that needs to be analyzed in terms of a set of sign systems, as figures 7.1 through 7.4 seek to make clear. Furthermore, referencing and citation behaviors vary within and between disciplines (i.e., they are field dependent), such that global criticisms of the phenomenon are misplaced. Perhaps the bibliometric baby need not be thrown out with the behavioral bath water, for, as Grafton (1997, 18) notes in his engaging history of the footnote, "a historical work and its notes can never, in the nature of things, reproduce or cite the full range of evidence they rely on," a view endorsed by Merton (1977, 84).

> *By itself*, citation analysis cannot trace all the complex sources of cognitive influences upon a particular work since *explicit citations*, which are ordinarily the only kind entered into quantitative citation-analyses, do not

adequately reflect the story. . . . A fine-grained analysis would have to be supplemented by focused interviews with scientists reporting on contexts of what they have set in print.

And even if such were possible to make visible all influences, explicit and tacit, who is to say with absolute authority that author *A* rather than *B* should have been cited at any given point? Who can say whether my reception of, negotiation, or interaction with another's text (reader response) is right or wrong, appropriate or inappropriate? What matters is that I have chosen to reference *A* rather than *B*. It is an affirmation, a form of voting (Fuller 1996; White 1990), which ought not to be judged on a scale of objective relevance. This raises the question of why authors vote the way they do and why certain signs are preferred over others. Helpful insights can be found within information retrieval (IR), where there has been much discussion of psychological relevance. Traditionally, relevance in IR has been defined in terms of topical coincidence—the extent to which two or more documents deal with the same subject(s). More recently, there has been a perceptible shift away from essentialist thinking, the idea that relevance should be defined exclusively in terms of topicality, what may be termed the Cranfield legacy (Harter and Hert 1998). For instance, Harter (1992) has argued that authors' perceptions of relevance within an IR context are often not defined exclusively in terms of topical correspondence. Building on the work of Sperber and Wilson (1986), he has hypothesized that our understanding of not only information seeking and retrieval but also referencing behaviors could be illuminated by drawing upon the theory of psychological relevance (Harter 1992, 613): "If psychological relevance is a factor in citation behavior, we would expect that at least some references made by citing articles would be to works that do not treat the topic of the citing articles."

Authors can be moved and their thinking shaped or redirected by a variety of cognitive, affective, and contextual factors that have little if anything to do with the ostensible subject matter of the citing paper; psychological relevance is, it could be argued, broader than topical relevance. At the very least, psychological factors may influence the choice of *A* over *B*, with both relevant, but *A* less than *B*. In an empirical study designed to test the degree of subject similarity between pairs of citing and cited documents (as reflected in the class numbers and descriptors assigned to them), Harter, Nisonger, and Weng (1993) discovered that the agreement level was frequently small, which, superficially, seemed to confirm the hypothesis that authors flag texts for reasons other than, or in addition to, topical affinity. They subsequently concluded that psychological relevance could be used to inform our understanding of bibliometrics. These findings raise questions about the assumptions and methods used by Mac-Roberts and MacRoberts (1986, 1989) to demonstrate nonrandom bias in citation practice. However, we are still left with a black-box explanation of citing behavior. Psychological relevance may help expose the limitations of the arguments used by the MacRoberts to challenge the validity of citation analysis, but it does not provide us with a particularly illuminating model of referencing dynamics.

Informational imperfections are a feature of everyday life. The fact that there are gaps in our knowledge of the motivational substrate does not require us to conclude that authors' referencing behaviors lack uniformity or are fundamentally haphazard. The weight of empirical evidence suggests that authors reference the works of their peers in a serious and normatively guided manner, and that these signs (references) perform a mutually intelligible communicative function. As van Braam (1991, 301) has demonstrated, giving "operational information" is the most important reason for referencing because of the interaction that takes place in the reviewing process between authors, editors, and referees. Either we have to assume that authors are engaged in repeated acts of whimsy, which just happen to be overlooked or go undetected by those responsible for quality control in the primary communication system—a curiously baroque hypothesis—or we have to conclude that the reasons for, and ways in, which authors invoke the works of other authors have become, in Merton's (1977, 48) words, "normatively operative in modern science."

To repeat a point made previously: if we accept that academic writing is a social act like a conversation, with rules for conversing (Brodkey 1987, 4), then the nuances with which these conversational cues and clues are imbued may be lost once they are wrested from their conversational scaffolding, like words repeated out of context and out of sequence by an uninformed third party. Czarniawska-Joerges (1998, 52), in fact, speaks of referencing as "a special mode of conversation with rules of its own." In the language of Bourdieu (1991, 20), citations are an instance of the "forms and formalities of a field which must be observed if discourse is to be produced successfully." Over the course of the last century, these literary inscriptions, examples of what Yates and Sumner (1997, 3) call "generically structured communications," have become indispensable features of any text with pretension of scholarly status and also taken on a shadow life as signs of esteem or prestige.

CONCEPT MARKERS/SYMBOLS

In order to understand fully the rhetoric of referencing, we need to examine the production processes and consumption practices associated with these signs, and how the meanings we attribute to them can vary under different conditions and also from discipline to discipline. More specifically, we should consider how a recently evolved craft practice (formal bibliographic referencing) has been effectively industrialized. The development of commercial citation indexes has given rise to the academic equivalent of the *Financial Times* or *Dow Jones Index*, from which the citation performance of blue-chip research institutions and "poster profs" can be charted and tracked, all part of today's "productivist scramble" (Wernick 1991, 169) within academia. The resultant league tables based on raw or weighted citation scores—indices of institutionalized prestige (Bourdieu 1991)—can now be generated on demand by ISI's research department, for a fee. Even if we never reach the point when an individualized personal citation summary is available at the touch of a button (Sosteric 1999), the trend to commodification is inescapable.

The central idea is very simple: collect, organize, and make available in machine readable form all bibliographic references from articles in reputable scholarly journals and allow users to search the database for those articles/authors which/who are pointed to with relatively high frequency on the not unreasonable assumption that frequency of mention, whether of authors or journals, has something to do with perceived utility or impact. But what happens to these signs (the references) when they are de-referentialized, flipped into citations, lifted from their natural settings, and homologized in the interests of convenience calculus— transmuted into the convertible currency of the citation? What, in other words, is lost, traduced, or falsely assumed in the translation, a point raised by Warner (1990, 28) in an early essay on the intersection of semiotics and information science: "the ambiguity of citations in aggregate form can be seen as a special case of the indeterminacy other written signifiers, such as words, can acquire when torn from their discursive context." His use of language in this extract is revealing: words are "torn from their discursive context." Warner seems not to grasp the fact that references and citations need to be unraveled in terms of their respective sign systems (figures 7.2, 7.4, and 7.5) if progress in terms of theory building is to be achieved.

Most certainly, references should not be dismissed as mere paratextual baubles or tools of persuasion (Gilbert 1977). Small (1978) has shown that in some subject domains—for instance, chemistry, where terminology is "hard"—the embedded reference has a clear referent. Specifically, he talks of citations as markers or symbols that denote "specific concepts or methods for particular disciplinary or speciality groups" (Small 1980, 187). On examining fifty of the most highly cited chemical papers, he found that 87 percent of the authors referencing those documents were citing them for the same reasons and using almost identical terminology when citing them. This suggests extremely tight coupling between signifier and signified, a view reinforced by van Raan (1998), rather like the idea of a pure performative, so transparent in intention that its meaning is unambiguous (Fish 1989, 44). But, as Small concedes, what holds for a selective population of experimental and theoretical chemistry articles, is unlikely to hold for polyparadigmatic fields, in which terminology, referencing styles, and worldviews may be much looser. As a general rule, the interpretant will be more or less clear-cut depending on the nature of the discipline and the role of references therein. For more on the subject of topic/concept mapping, see the excellent review by White (2004).

READER RESPONSE

The residual subjectivity of referencing behavior raises questions relating to the construct validity of citation analysis and evaluative bibliometrics. For a vocal minority within the academic community, citation counting and ranking amount to little more than numerology; the citation has been "fetishized and turned into a highly desirable and marketable commodity" (Hicks and Potter 1991, 483). This view was also aired by Apostol (1998), whose spirited caricature of bibliometric bean counting appeared in *Science Tribune* (see Cronin [1998b] for a rejoinder).

> Bibliometrics is a numerics of papers, citations, and other items which are sometimes related, but accidentally, to the substance of scientific research. Bibliometricians count all day long: they count how many papers an author has published, in how many journals, in how many ways; they count affiliations, acknowledgments references; they count figures, tables, equations, words, key words, paragraphs, etc., etc. . . . After counting, they compute, add, subtract, divide, multiply, exponentiate, take logarithms and, even sometimes, do more sophisticated analyses.

Without reliable knowledge of why an author references the work of another, assumptions about the full communicative import of such signs are open to question, for, as Deetz (1994, 212) notes, "[h]uman beings are interpretative creatures." That, necessarily, raises concerns as to the validity of treating all citations as unitary equivalents that can be bundled together and statistically manipulated to make judgments about the relative research performance of individual scholars, project teams, institutions, or nation states. However, the *reductio ad absurdum* of the radical interpretivist position is to deny outright the sociometric and information theoretic significance of collective referencing behaviors. That, in my opinion, is too high a price to pay. Personal and institutional investors trade stocks on the basis of less than perfect information; why should speculation in scholarly futures, distasteful though the idea may be to some, operate any differently in principle, given what we know about the pecuniary correlatives of citations?

Reader response theory offers an alternative perspective. A central problem with citation analysis is held to be the unknowability and inarticulability of the motivations that shape authors' sign actions. Small's (1978, 1980) work notwithstanding, there are also legitimate questions to be raised regarding the meaning readers extract from references—the flip side of the interpretative coin. Granted, certain elements of the sign have formal or objective properties (the cited author's name, journal title, date of publication, publisher, broad subject matter, etc.), but readers, even if belonging to the same discourse community, may decode aspects of these signs, and their referents, differently—the "mirage of the referent," in Baudrillard's (1981, 150) pert phrase. The possibility of noise arising at both the sender end (multiple motivations) and receiver

end (multiple meanings) of the communication spectrum is real. But
that does not require us to conclude that all references are, to paraphrase
Fish (1989, 44), "equally and radically orphaned in the sense that no
one of them is securely fastened to an independently specifiable state of
affairs." And it certainly has not deterred proponents of citation analysis from investing these signs with considerable symbolic value.

THE CYBER-PANOPTICON

As indicators of professional esteem, citations are highly valued in today's academic marketplace. The shift from text to individual as the unit
of analysis creates the conditions necessary for the establishment of a
symbolic capital market (compare figures 7.4 and 7.5). This is a non-
trivial sign difference, one that is often misunderstood. Sosteric (1999)
notes that citation indices can be used to make the work of scholars visible to administrators of the scientific enterprise. For rankings-conscious
deans, provosts, or presidents, citation analysis offers the enticing
prospect of being able to check the performance of different research
groups and highlight centers of excellence (Moed et al. 1985), while
journal editors and their commercial publishers are becoming increasingly sensitized to the impact factors of their journals despite validity
concerns (Seglen 1997; Tobin 2003). What we are witnessing, to appropriate the language of Baudrillard (1981, 92), is "the sanctification of the
system as such, of the commodity as system."

While some may welcome the transparency-inducing effects of citation analysis, others prefer to see it as evidence of an emerging culture
of cybernetic control within academia. Sosteric (1999) construes citation analysis as a form of "Orwellian surveillance net" that can be used
to generate performance data on individual faculty members virtually
on demand. The Jeremiahs will surely claim that the age of electronic
panopticism has dawned in academe.

> We have drawn on Foucault . . . because we were interested in the paral-
> lels between Foucault's analysis of the development of institutions such
> as the penal system and medicine as new technologies of observation and
> control, and the work of citation analysis (CA) itself, which provides a
> new way of making scientific practice visible and recordable, and new
> possibilities for producing hierarchies of difference and categories of nor-
> mal/abnormal scientific behavior. (Hicks and Potter 1991, 475)

This statement, however, misses the point that such developments are an integral feature of advanced societies—and not necessarily detrimental to individual welfare (Webster 1995, 69). The meanings of references have little to do with the significance accorded to citations in the context of programmatic assessments. Evaluative bibliometrics ignores the grounded or situated nature of references, treating them instead as decontextualized citations; as indicators of a scholar's/researcher's worth, merit, or esteem, along with other objectified signs (e.g., publication counts, grant awards, scientific honors). The referent of the bibliographic reference is a specific work; the referent of a citation the absent text that it denotes (and all the other texts that point to it); in the case of large-scale citation counts, the referents are the cited authors (rather than their published works). The need to disentangle these three sign systems (see, once again, figures 7.2, 7.4, and 7.5) is clearly exemplified in the context of the British higher education system, which affords a useful case study.

In the United Kingdom, the prospect of institutionalized citation analysis, though slim, is arguably stronger than in almost any other country, and it is worth looking in some detail at how the British government uses performance indicators (summable signs of scholarship) to foster selectivity, sometimes against the expressed wishes of the academic establishment (Anderson 1991). Although antipathy toward evaluative bibliometrics has a long history in the United Kingdom, it is not hard to understand the appeal of a putatively objective performance assessment tool, such as citation analysis, within a powerfully centralized higher education system (see, for example, Healey, Irvine, and Martin 1988). Dirigisme has become the order of the day, routinely manifested in regular research assessment exercises (RAEs) and teaching quality audits, the ground rules for which are notoriously fungible. In a matter of a decade or two, the balance of power in British higher education has been altered irrevocably, a point captured in the title of Halsey's (1995) expansive survey of recent trends, *Decline of Donnish Dominion*. With funding tied to national research rankings (themselves derived from heterogeneous sign systems), competition among universities has reached unprecedented levels.

Britain may have been the birth-place of soccer, but few would have predicted the extent to which the tactics of soccer managers have taken over the world of higher education. Last month the government announced the results of its research assessment exercise: a league table of departments and institutions resulting from the world's most comprehensive peer-review process. . . . This year's results revealed how soccer-style transfers of researchers and other tactics aimed at improving a department's rating are now part of British academic life. (Williams 1997, 18)

Such developments have ruptured established cultural practices within the British university system.

SIGN AS SYNECDOCHE

The soccer analogy can be extended, with publications and citations being the equivalent of goals scored, on the basis of which bonuses are paid out to high-performing stars or teams. Quality within the context of the RAE was originally calibrated on a 1–5 nominal scale. The poles translated into "national excellence in none or very few areas of activity" and "international excellence in some areas of activity and national excellence in the remainder," respectively. Expert review panels determined departmental ratings for different subject areas, using both quantitative and qualitative indicators. These included peer assessment, number of research students, and per capita performance indicators such as authored books, refereed articles, and research income. Currently, citation counts are not a component of the RAE model, though the studies by Oppenheim (1995, 1997) make a superficially cogent case for using them as proxies for other more costly or labor-intensive performance measures. This takes us well beyond the semiotics of referencing to the commodification and ultimate celebration of symbols of esteem.

In the event of the Higher Education Funding Councils (HEFC) electing to use citation counts as the sole or ultimate arbiter of departmental performance in some future research assessment exercise, we would by then have reached the point of sign as synecdoche. In such a scenario, an inevitable concern would have to do with the extent to which the multidimensionality of scholarly inquiry and productivity could be captured credibly through citation counts (for contrasting perspectives, see Baddeley [1998] and Lachmann and Rowlinson [1997]). More specifically, if an individual's, department's, or university's ability to amass symbolic capital of this kind were to become the critical determinant of future research funding and career advancement, then it would not be difficult to imagine distortions creeping into the system, as players devised recruitment, publication, collaboration, and citation harvesting stratagems to maximize the accumulation of symbolic capital. The transformation of these stockpiles of manipulated and manipulable capital into objectified ratings and rankings would, in turn, create a most convenient index for academic administrators to justify selective programmatic and institutional investment, what Sosteric (1999, 17) seems to have been implying by the phrase "the *sensing* methodology (or device) of a cybernetic system." The academy would finally have, in the words of Foucault (1997, 208) commenting on the new physics of power represented by panopticism, "mechanisms that analyze distributions, gaps, series, combinations, and which use instruments that render visible, record, differentiate and compare."

CONCLUSIONS

What semiotics offers information scientists and others engaged in bibliometric research is a supradisciplinary suite of insights and exegetical tools (such as the sign triads in figures 7.1 through 7.5) to better explore the indexical significance of bibliographic references and citations—both contextualized and decontextualized. Commercial citation indices have liberated references from their textual hosts, in the process creating a marketplace for a new species of sign, the citation. An understanding of semiotic principles may be one way of helping the broader scientometrics research community develop greater sensitivity to the variable symbolic significance of the signs they routinely manipulate and treat as quasi-objective indicators of quality, impact, and esteem. Semiotics cannot provide a unifying theoretical framework for understanding the intentional and extensional significance of citation, but it does at least offer a flexible scaffolding within which to examine specific phenomena and reproducible practices and to assess the strengths and limitations of the competing theoretical models.

REFERENCES

Anderson, A. (1991). No citation analyses please, we're British. *Science*, 252(5006), 639.

Apostol, M. (1998). Science against science. *Science Tribune*, Article—April 1998. Available at www.tribunes.com/ tribune/art98/apo2.htm.

Baddeley, A. (1998, June). So where should we publish? *Psychologist*, 312.

Baudrillard, J. (1981). *For a Critique of the Political Economy of the Sign*. London: Telos Press.

Ben-Ari, E. (1987). On acknowledgements in ethnographies. *Journal of Anthropological Research*, 43(1), 63–64.

Bourdieu, P. (1986). The forms of capital. In: Richardson, J. (Ed.). *Handbook of Theory and Research for the Sociology of Education*. New York: Greenwood Press, 241–258.

Bourdieu, P. (1990). *Homo Academicus*. Cambridge: Polity Press.

Bourdieu, P. (1991). *Language and Symbolic Power*. Thompson, J. B. (Ed.). Cambridge, MA: Harvard University Press.

Bourdieu, P., Passeron, J.-C., and Saint Martin, J. de. (1984). *Academic Discourse: Linguistic Misunderstanding and Professorial Power*. London: Polity Press.

Brier, S. (2004). Cybersemiotics and the problems of the information-processing paradigm as a candidate for a unified science of information behind library information science. *Library Trends*, 52(3), 629–657.

Brodkey, L. (1987). *Academic Writing as Social Practice*. Philadelphia: Temple University Press.

Budd, J. M. (2000). Scholarly productivity of U.S. LIS faculty: An update. *Library Quarterly*, 70(2), 230–245.

Budd, J. M. (2001). Instances of ideology in discursive practice: Implications for library and information science. *Library Quarterly*, 71(4), 498–517.

Burris, V. (2004). The academic caste system; Prestige hierarchies in PhD exchange networks. *American Sociological Review*, 69, 239–264.

Collini, S. (1989). Publish—and be dimmed. *Times Higher Education Supplement*, March 2, 13–14.

Cronin, B. (1981). The need for a theory of citing. *Journal of Documentation*, 37(1), 16–24.

Cronin, B. (1984). *The Citation Process: The Role and Significance of Citations in Scientific Communication*. London: Taylor Graham.

Cronin, B. (1994). Tiered citation and measures of document similarity. *Journal of the American Society for Information Science*, 45(7), 537–538.

Cronin, B. (1995). *The Scholar's Courtesy: The Role of Acknowledgement in the Primary Communication Process*. London: Taylor Graham.

Cronin, B. (1996). Rates of return to citation. *Journal of Documentation*, 52(2), 188–197.

Cronin, B. (1998a). Metatheorizing citation. *Scientometrics*, 43(1), 45–55.

Cronin, B. (1998b, June). New age numerology: A gloss on Apostol. *Science Tribune*, Commentary. Available at www.tribunes.com/tribune/art98/cron.htm.

Cronin, B. (1999). The Warholian moment and other proto-indicators of scholarly salience. *Journal of the American Society for Information Science*, 50(10), 953–955.

Cronin, B. (2000). Semiotics and evaluative bibliometrics *Journal of Documentation*, 56(3), 440–453.

Cronin, B. and Shaw, D. (2002). Banking (on) different forms of symbolic capital. *Journal of the American Society for Information Science and Technology*, 53(13), 1267–1270.

Cronin, B. and Weaver-Wozniak, S. (1993). Online access to acknowledgements. In: Williams, M. E. (Ed.). *14th National Online Meetings: Proceedings—1993, New York, May 4–6*. Medford, NJ: Learned Information, 93–98.

Cronin, B., Snyder, H. W., Rosenbaum, H., Martinson, A., and Callahan, E. (1998). Invoked on the Web. *Journal of the American Society for Information Science*, 49(14), 1319–1328.

Czarniawska-Joerges, B. (1998). *Narrative Approach to Organization Studies*. London: Sage.

Davenport, E. and Cronin, B. (2001). The citation network as a prototype for representing trust in virtual environments. In: Cronin, B. and Atkins, H. B. (Eds.). *The Web of Knowledge: A Festschrift in Honor of Eugene Garfield*. Medford, NJ: Information Today, 517–534.

Deetz, S. A. (1994). Representational practices and the political analysis of corporations: Building a communication perspective in organizational studies. In: Kovacic, B. (Ed.). *New Approaches to Organizational Communication*. Albany: SUNY Press, 211–244.

Edge, D. (1977). Why I am not a co-citationist. *Society for the Social Studies of Science Newsletter*, 2, 13–19.

Egghe, L. and Rousseau, R. (1990). *Introduction to Informetrics: Quantitative Methods in Library, Documentation and Information Science*. Amsterdam: Elsevier.

Fish, S. (1989). *Doing What Comes Naturally: Change, Rhetoric, and the Practice of Theory in Literary and Legal Studies*. Durham, NC: Duke University Press.

Foucault, M. (1997). *Discipline and Punish: The Birth of the Prison*. New York: Random House.

Fuller, S. (1996). Personal communication.

Garfield, E. (1997). Letter to the Editor. *Journal of the American Society for Information Science*, 48(10), 926.

Garfield, E. (1998). I had a dream . . . about uncitedness. *Scientist*, 12(14), 10.

Genette, G. (1997). *Palimpsests: Literature in the Second Degree*. Lincoln, NB: University of Nebraska Press.

Gilbert, G. N. (1977). Referencing as persuasion. *Social Studies of Science*, 7, 113–122.

Gluck, M. (1997). Making sense of semiotics: Privileging respondents in revealing contextual geographic syntactic and semantic codes. In: Vakkari, P., Savolainen, R., and Dervin, B. (Eds.). *Information Seeking in Context. Proceedings of an International Conference on Research in Information Needs, Seeking, and Use in Different Contexts, 14–16 August, 1996*, Tampere, Finland. London: Taylor Graham, 53–66.

Grafton, A. (1997). *The Footnote: A Curious History.* Cambridge, MA: Harvard University Press.

Halsey, A. H. (1995). *Decline of Donnish Dominion: The British Academic Professions in the Twentieth Century.* Oxford: Oxford University Press.

Harter, S. P. (1992). Psychological relevance and information science. *Journal of the American Society for Information Science*, 43(9), 602–615.

Harter, S. P. and Hert, C. A. (1998). Information retrieval evaluation: Toward a unified framework. In: M. E. Williams (Ed.). *Annual Review of Information Science and Technology*, 33, Medford, NJ: Information Today, 3–94

Harter, S. P., Nisonger, T. E., and Weng, A. (1993). Semantic relationships between cited and citing articles in library and information science journals. *Journal of the American Society for Information Science*, 44(9), 543–552.

Healey, P., Irvine, J., and Martin, B. R. (1988). Introduction: Quantitative science-policy studies in the United Kingdom. *Scientometrics*, 14(3–4), 177–183.

Hicks, D. and Potter, J. (1991). Sociology of scientific knowledge: A reflexive citation analysis *or* science disciplines and disciplining science. *Social Studies of Science*, 21, 459–501.

Hyland, K. (2000). *Disciplinary Discourses: Social Interaction in Academic Writing.* Harlow: Pearson Education.

Lachmann, P. and Rowlinson, J. (1997, Winter). It's not where you publish that matters. *Science and Public Affairs*, 8.

Landes, W. M. and Posner, R. A. (2000). Citations, age, fame, and the Web. *Journal of Legal Studies*, 29(319), 329–341.

Leydesdorff, L. (1998). Comments on theories of citation? *Scientometrics*, 43(1), 5–25.

Liu, X. (1993). The complexities of citation practice: A review of citation studies. *Journal of Documentation*, 49(4), 370–408.

MacRoberts, M. (1997). Letter to the Editor. *Journal of the American Society for Information Science*, 48(10), 963.

MacRoberts, M. H. and MacRoberts, B. R. (1986). Quantitative measures of communication in science: A study of the formal level. *Social Studies of Science*, 16, 151–172.

MacRoberts, M. H. and MacRoberts, B. R. (1989). Problems of citation analysis: A critical review. *Journal of the American Society for Information Science*, 40(5), 342–349.

McCain, K. W. (1991). Communication, competition, and secrecy: The production and dissemination of research-related information in genetics. *Science, Technology, & Human Values*, 16(4), 491–516.

Merton, R. K. (1977). The sociology of science: an episodic memoir. In: Merton, R.K. and Gaston, J. (Eds.). *The Sociology of Science in Europe*. Carbondale: Southern Illinois University Press, 3–141.

Moed, H. F., Burger, W. J. M., Frankfort, J. G., and van Raan, A. F. J. (1985). The use of bibliometric data for the measurement of university research performance. *Research Policy*, 14, 131–149.

Nelson, C. (1997). Superstars. *Academe*, 87(1), 38–54.

Oppenheim, C. (1995). The correlation between citation counts and the 1992 Research Assessment Exercise ratings for British Library and Information Science university departments. *Journal of Documentation*, 51(1), 18–27.

Oppenheim, C. (1997). The correlation between citation counts and the 1992 research assessment exercise ratings for British research in genetics, anatomy and archaeology. *Journal of Documentation*, 53(5), 477–487.

Posner, R. A. (2001). *Public Intellectuals: A Study of Decline*. Cambridge, MA: Harvard University Press.

Price, D. J. de S. (1963). *Little Science, Big Science*. New York: Columbia University Press.

Rotman, B. (1987). *Signifying Nothing: The Semiotics of Zero*. New York: St. Martin's Press.

Rousseau, R. (1997). Sitations: an exploratory study. *Cybermetrics*, 1(1). Available at www.cindoc.csic.es/cybermetrics/articles/v1p1html.

Sebeok, T. A. (1994). The study of signs. In: Sebeok, T. A. *Signs: An Introduction to Semiotics*. Toronto: University of Toronto Press, 3–10.

Seglen, P. O. (1997). Why the impact factor of journals should not be used for evaluating research. *British Medical Journal*, 314. Available at http://bmj.bmjjournals.com/cgi/content/full/314/7079/497.

Small, H. (1978). Cited documents as concept symbols. *Social Studies of Science*, 8, 327–340.

Small H. (1980). Co-citation context analysis and the structure of paradigms. *Journal of Documentation*, 36(3), 183–196.

Sosteric, M. (1999). Endowing mediocrity: Neoliberalism, information technology, and the decline of radical pedagogy. *Radical Pedagogy*, 1(1). Available at www.icaap.org/iuicode?2.1.1.3.

Sperber, D. and Wilson, D. (1986). *Relevance: Communication and Cognition*. Cambridge, MA: Harvard University Press.

Spurgin, T. (2001). The *Times Magazine* and the academic megastars. *Minnesota Review*, 52–54, 225–238.

Tobin, M. J. (2003). Editorial. Impact factor and the *Journal*. *American Journal of Respiratory and Critical Care Medicine*, 168, 621–622.

Trow, M. (1998). On the accountability of higher education in the United States. In: Bowen, W. G. and Shapiro, H. T. (Eds.). *Universities and Their Leadership*. Princeton, NJ: Princeton University Press, 15–61.

van Braam, R. R. (1991). *Mapping of Science: Foci of Intellectual Interest in Scientific Literature*. Leiden: University of Leiden Press/DSWO Press.

van Raan, A. F. J. (1998). In matters quantitative studies of science the fault of theorists is offering too little and asking too much. *Scientometrics*, 43(1), 129–139.

Warner, J. (1990). Semiotics, information science, documents and computers. *Journal of Documentation*, 46(1), 16–32.

Webster, F. (1995). *Theories of the Information Society*. London: Routledge.

Wernick, A. (1991). *Promotional Culture*. Newbury Park, CA: Sage.

White, H. D. (1990). Author co-citation analysis: Overview and defense. In: Borgman, C. L. (Ed.). *Scholarly Communication and Bibliometrics*. Newbury Park, CA: Sage, 84–106.

White, H. D. (2004). Citation analysis and discourse analysis revisited. *Applied Linguistics*, 25(1), 89–116.

Williams, N. (1997). U.K. universities: jostling for rank. *Science*, 275, January 3, 18–19.

Wouters, P. (1993). Writing histories of scientometrics or what precisely is scientometrics? Available from the author.

Wouters, P. (1998). The signs of science. *Scientometrics*, 41(1–2), 225–241.

Yates, S. J. and Sumner, T. R. (1997). Digital genres and the new burden of fixity. In: Sprague, R. H. (Ed.). *Proceedings of the Thirtieth Hawaii International Conference on System Sciences*, Vol. VI. Los Alamitos, CA: IEEE Computer Society Press, 3–12.

Chapter Eight

The Attention Economy

Why do we cite? The answer, naturally, will depend on the perspective one favors and the degree of granularity required. I present three perspectives (briefly considered in the previous chapter), which I have labeled the functionalist, normative, and phenomenological. Studies of citation practice typically approach the topic from a single perspective. Intuitively, that makes little sense. Missing from the literature are compelling attempts to achieve discourse synthesis between different investigative communities and worldviews, with the result that the citation field, like many others—for instance, sociology (Turner 1989)—is characterized by logically connected but noninteracting literature sets (Swanson 1989). In a wide-ranging review of citation studies, Hemlin (1996, 227) observes that the "mixture of results and methods does not lead to a proposal for a theory of citation behavior," while Hicks and Potter (1991, 480), adopting a Foucaultian stance, suggest that "the act of citation is a symptom of many . . . things." The quest for *a* theory of citation (Cronin 1981) may well be a misguided undertaking (Leydesdorff 1998).

A purely functionalist interpretation of why *A* cites a text (T1) by *B* might run as follows: to provide supplementary evidence, to support or refute a hypothesis, or to furnish historical context. Of course, what is cited can be as simple as a specific term or as complex as an author's complete oeuvre (Cronin 1994). Various schemata and relational operators/qualifiers have been devised to accommodate the spectrum of ostensible reasons for referencing a particular text (or cognitive resource), but none comprehensively answers the question, "Why do authors cite?" Van Braam (1991, 301) argues that, "because of the interaction that takes place in the reviewing process between authors, editors and referees," citations cannot be dismissed as "private symbols." In similar vein, Small (1978) has argued that in the physical (and life) sciences citations are tightly coupled (he speaks of citations as concept symbols) with specific experimental designs, analytic techniques, observable elements and their properties, reaction times, interaction effects, statistical methods, and such like (see also White [2004]). That, of course, is very different from established practices in the social scientific, humanistic, and belletristic literatures. And there is the historical dimension: in seventeenth-century science, authentication of experimental practice and results was provided by individuals of gentlemanly stock (Shapin 1994). The social trust implicit in this civil process of verification by a third party finds its echo today in the depersonalized practice of citation. Pragmatically speaking, citations bear witness to earlier scientific events and outcomes, and serve a serious rather than a purely rhetorical or decorative purpose in the narrating of science.

A normative, or domain analytic (Hjørland 1997), interpretation would highlight the rules, tacit or codified, that govern the dispensing of credits (rewards) within the scholarly communication system and its constituent domains. As Bourdieu (1991, 20) notes, "if one wishes to produce discourse successfully within a particular field, one must observe the forms and formalities of that field." This includes matters of etiquette and style, as well as more generalized scholastic savoir faire. Typically, these skills are acquired through an admixture of osmosis, formal and informal mentoring, professional communication, and interaction (Brodkey 1987; Czarniawska 1997) and the quotidian transmission of cultural precedent from master to apprentice, peer to peer—"the Apostolic succession of apprenticeship" (Turner 1994, 50).

But neither the functionalist nor normative explanation fully answers the question, why did *A* cite T1 by *B*? There remains the motivational dimension. How are citations used to persuade the reader or provide rhetorical buttressing for the author's thesis? Why was T2 by author *C* not cited? What motivated *A* to cite T1 rather than any other candidate text, whether written by *B* or another? What social-psychological variables come into play in the shaping of an author's citation behavior? And if these are not known (or knowable), how can a putative (and plausible) theory of citation be developed? How, then, to use the language of economists, are these informational imperfections to be overcome? The Achilles' heel of citation is its residual subjectivity, and this has spurred MacRoberts (1997) to challenge the validity of evaluative bibliometrics on the grounds that authors' citation behavior is nonrandomly (i.e., systematically) biased. But absolutism is ill advised, since it is unreasonable to expect authors, no matter how conscientious, to identify and cite all potentially relevant earlier works (see Grafton 1997, 18). Some degree of selection bias is inevitable.

TOWARD A UNITARY PERSPECTIVE

These three perspectives (functionalist, normative, phenomenological) are not mutually exclusive; citations have multiple articulations, and citation analysis, as Leydesdoff (1998) maintains, allows for movement between the cognitive, textual, and social dimensions of science. For example, *A*'s citing of *B*'s text may provide supporting evidence for *A*'s theory (the citation's perceived intratextual function), while his formal citing of *B*'s work reflects the scholarly community's expectation that credit will be given to those who have influenced the citing author's thinking (its extratextual function). The citation, in that sense, is socially mandated. But the specific reasons for citing *B* rather than any other author may also be contingent upon a variety of social, structural, cultural, economic, or organizational factors—"the hypothetical causal chain" (Shadish 1989, 408). Thus, a full understanding of why *A* cites *B* requires a multilayered explanation and, ideally, thick description of the process—and the politics of the process. Most citation classification schemes, however, suffer from an understandable desire to uniquely pigeonhole authorial purpose or intention, thus downplaying the mix of normative and situational factors at work.

In addition to any overtly instrumental role they may play within a given text, citations can act as signaling or strategizing devices (Ben-Ari 1987) and also as units of potentially convertible symbolic capital in the academic marketplace. The nature of the interaction between these symbolic entities (citations) and related social processes (scholars' information consumption and communication practices) are not readily apprehensible. Absent a unified theory of citation, a range of metaphors has been used to fill the gap. Becher (1989) talks of citation in terms of levying a tax on reusable knowledge; I (Cronin 1981, 16) have resorted to the image of frozen footprints in the landscape of scholarly achievement; Hagstrom (1982) sees citation in the context of gift giving (the bestowing of social recognition on one's colleagues in exchange for information). This last interpretation has been roundly dismissed by Latour and Woolgar (1982, 38) as having "the aura of a rather contrived fairy tale."

SHADOWLANDS SOCIOLOGY

Prevailing citation practices are of fairly recent origin, as Leydesdorff (1998) recognizes, and are a consequence of the progressive professionalization and institutionalization of the scientific enterprise. However, citations are anything but trivial literary inscriptions, or instances of reflexive ritual in the production of stylized texts; the sanctions for failure to cite others' work can be severe and the career costs considerable. Citation has become a structurally embedded component of the primary communication process, and thus deserves to be included in epistemological critiques of science. Cooperation and collaboration are also defining features of the scholarly enterprise, illustrated persuasively in the growth of networking, coauthorship, and informal communication in recent decades. The nature of the linkages that define disciplinary and social networks in the various sciences is revealed through webs of citations and acknowledgments.

Acknowledgments, though they register essentially personal or behind-the-scenes interactions, gift giving, and informal know-how trading and mentoring—which includes the contributions of "silent scientists" (Meadows 1974, 182)—do, nonetheless, exist in the public domain, and when analyzed in sufficient numbers provide a revealing window onto trends in collaboration, particularly if used conjointly with other measures of scholarly interdependence and interaction (Giles and Councill 2004). One might use the term "biographic coupling," playing on the idea of bibliographic coupling, to convey the fact that two authors have been coacknowledged one or more times in a given corpus of documents.

However, this operationalization of trusted assessorship (Chubin 1975; Mullins 1973) has been largely neglected in the interpretative sociology of science. Unlike citations, acknowledgments have been relegated, historically, to the shadowlands of scholarly production, but the case for their "artful integration," to paraphrase Suchman (1996, 407), into mainstream accounts of science and scholarship is increasingly difficult to resist in the light of the empirical evidence (Cronin 1995; McCain 1991). Leydesdorff (1998) really needs to extend his focus to accommodate other generic forms of citation—acknowledgments, footnotes, and dedications. When measured and mapped, these complementary, paratextual elements take on a significance that transcends the superficially ritualistic roles they perform in the process of academic writing. Carefully aggregated and combined, these paratextualities constitute a robust, composite indicator of scholarly interaction, impact, and perceived utility.

How might the many strands of the citation debate be woven into a coherent whole? Leydesdorff (1998) argues that the quest for a grand theory of citation implies a metatheoretical question. That being so, two (rather different) approaches are proposed here for consideration. Metatheories, to quote Vakkari (1997, 452), do not "signify substantive theories" and "are not about particular social structures, processes or groupings"; instead they provide a flexible scaffolding or framework within which to think about and explain specific sociological phenomena (Berger, Wagner, and Zelditch 1989; Vakkari and Kuokkanem 1997; Vickery 1997). The first approach involves a recontextualizing of citation practices to accommodate the interplay of the political and the personal in the production and exploitation of symbolic capital; it is recommended that citation behavior be examined in the context of a political and moral economy. The second approach would apply structuration theory to citation behavior in an effort to bridge the existing interpretive divide and to articulate the relationship between private acts and public worlds. Both of these metatheoretical approaches are served up here in extremely raw form.

THE MORAL AND POLITICAL ECONOMY OF CITATION

The multidimensionality, or multicontextuality, of citation analysis has been well captured in Leydesdorff's (1998) two-by-two tables, which reveal the "functions of citation relations" at both the micro/disaggregated and macro/aggregated levels. What these tables so clearly show is that a number of socioprofessional relations are established (between actors, texts, and cognitive resources) when *A* cites *B* cites *C*, etc. When the unit of analysis is clusters of authors, we are dealing with networks of texts, cognitive resources, and social actors—and networks of networks. The interpretation of these networks requires an appreciation of the relationships, power structures, alliances, institutional affiliations, and bases of prestige that animate the moral and political economy of citation.

Bibliometric and sociometric analyses of specialty groups are predicated on assumptions about the purpose and integrity of individuals' citation behaviors. The outputs of such studies metastasize into maps of science and information theoretic models of disciplines. Citation analysis makes invisible colleges and virtual communities manifest; it effectively provides an explanation for the growth of science and how influence is exercised within and across scientific clusters. In effect, citation analysis has morphed into a tool for describing in shorthand the growth of contemporary science, for calibrating academic performance, and for allocating credits and rewards. Despite persistent criticisms relating to construct validity and reliability (MacRoberts and MacRoberts 1989), citation analysis remains, according to the weight of empirical evidence, an efficient means of measuring, if not perceived quality, then perceived usefulness: "articles are highly cited if they are useful to a large number of scientists" (Shadish 1989, 415). The net effect is that citations—to resort to phrasing used earlier—have become the "most objectified of the indices of symbolic capital" (Bourdieu 1990, 76), in part as a result of their endorsement by a variety of science policy and research funding agencies. A corollary of this is the growing recognition that market signaling, linked to a publication/citation maximization strategy, whether consciously applied or not, can translate into significant economic gains for the perpetrator (Cronin 1996). A nagging difficulty, however, has to do with the fact that the process (that which is signified by the act of citation) is divorced from its context once the signifier is subjected to number crunching, a point made by Warner (1990, 28). However, the anti-commodificationists are still left to wrestle with the pragmatic realism of White (1990, 91), namely, that citations are like votes that have been cast, and when cast in large numbers they tell us which authors and works have had an impact. The fact that we know little or nothing about voter's motivations is neither here nor there.

How might these contesting viewpoints be accommodated? One way is to think in terms of the moral economy of citation, to appropriate Silverstone, Hirsch, and Morley's (1992) term. By this I mean to refer to the inherent mutuality of science, with shared expectations of granting credits and acknowledging debts in a transactional system. Transposing Silverstone et al.'s (1992, 18) definition of what constitutes a moral economy of the household, one could say that authors' citation practices are "defined and informed by a set of cognitions, evaluations, and aesthetics, which are themselves defined and informed by the histories, biographies and politics" of the scholarly communication system and that citations are "doubly articulated into public and private culture" (1992, 15). Indeed, the sense that authors have a collective understanding of the rationale for, and specifics of, citation conjures up the idea of collective mind, conceptualized, in the context of organizational studies, by Weick and Roberts (1993, 357) as "a pattern of heedful interrelations of actions in a social system." Citation is predicated on assumptions about socially correct and acceptable practice—distributive justice, in other words. Acknowledging one's intellectual debts and paying credit through bestowal of citations is a means of positioning and patterning oneself vis-à-vis the external community or networks of which one is a member.

STRUCTURATION

The problematic relationship between structure and action has been an important topic in twentieth-century sociology. Giddens (1984) has attempted to bridge the theoretical divide via structuration theory, which "subsumes two fundamentally antagonistic theoretical positions, that of the structuralist who sees social life as determined by objective social structures and that of the hermeneutical humanists and interactionists who see social life as a product of subjective and intersubjective human activity" (Macintosh and Scapens 1990, 456). In structuration terms, social practices, defined as "the concrete, situated, and routine activities in which people are engaged as they enter, work, play in, and exit a variety of social settings" (Rosenbaum 1996, 86), are routinely reproduced in social settings by people; social actors in this formulation are individuals—not organizations, collectivities, systems or social structures. And they combine discursive knowledge

(which can be articulated) with practical, or tacit, knowledge that re-
sists articulation. Structure, in turn, is defined in terms of the rule sets
(techniques or generalizable procedures) and resources (facilities or
bases of power) that are invoked and used in the production and reproduction of social systems.

In the literature on citation there exists a comparable ideological cleft
between the normative and interpretivist positions (Cronin 1984). Simply put, the normative view posits a world in which citation behaviors
are rational, universalistic, and rule based. Interpretivists, on the other
hand, eschew sweeping generalizations and normative accounts of science, privileging instead "the deep personal character of science"
(Mitroff 1974, 580). The difference could equally well be formulated as
objectivists versus subjectivists, or those who are comfortable with the
institutionalization of citation analysis and those who dismiss the practice as nothing more than "numerology" (Jevons 1973, 45).

Can structuration theory provide a means of bridging the citational
cleavage described by Leydesdorff (1998)? Citation is a well-established
practice, rooted in a recognizable social system, that of scholarly communication. Scholars have a mixture of discursive and practical knowledge about citation (the why, how, and wherefor) and, in the process of
producing and disseminating their publications, routinely draw upon
cognitive resources (the ideas, insights, and texts of other actors) and ad-
here to rules (governing the allocation of credits, distribution of symbolic and material rewards). These rules, following Giddens, are key in
the reproduction of the institutionalized social practice known as citation
behavior. Structuration may thus offer the possibility of carving out a
middle ground between naïve normativism and extreme relativism.

A RETROSPECT

Since the idea of a unified citation index to the literature of science was
first outlined by Garfield (1955), the *Science Citation Index* has established itself as the gold standard for scientific information retrieval. It
has also become the database of choice for citation analysts and evaluative bibliometricians worldwide. As scientific publication moves to the
web, and novel approaches to scholarly communication and peer review
establish themselves, new methods of citation and link analysis are be-

ing developed to capture often liminal expressions of peer esteem, influence and approbation. The web thus affords bibliometricians rich opportunities to apply and adapt their techniques to new contexts and content; the age of "bibliometric spectroscopy"(van Raan 2000, 309) has dawned.

Bibliometricians count and measure things. Traditionally, they have concentrated their efforts on tracking highly visible and objective indicators of scholarly activity, most notably publications and citations. Bibliometric techniques can, of course, be applied to many other inputs, outputs, processes, and artifacts associated with the conduct of science (journals, acknowledgments, scientific manpower, federal funding patterns, rates of patenting, etc.), in which case either "informetrics" or "scientometrics" may be a more apposite descriptor. However, the models and methods used by these intersecting research communities have much in common, and some of the subtle procedural distinctions may well blur as all three groups strive to develop context-sensitive methodologies and metrics appropriate to web environments.

Bibliometric analysis predates the development of the *SCI*, but the advent of the *SCI*, and specifically the availability of electronic access (online, CD-ROM, and web-based) to the Institute for Scientific Information's massive datasets, has had a catalytic effect on the popularity, scope, and ambition of bibliometric research, both within and also well beyond the information science community. The *SCI* grew out of a specialty index to the literature of genetics, and was, as Garfield readily concedes, inspired by Shephard's legal citation index, which was itself created almost a century earlier (Wouters 1999). Others, in fact, have argued that citation indexing's true conceptual origins are to be found in fourteenth-century Hebrew literature (Weinberg 1997). The *SCI* was conceived as a practical tool for indexing and retrieving the literature of science, but it was not long before the full import and ramified significance of Garfield's invention began to be appreciated by scholars in a number of disciplines (Merton 2000). Subsequent development of sophisticated techniques for representing and mapping networks of scientific papers and authors (e.g., cocitation analysis) have taken us well beyond the principles and practice of information retrieval, the commercial raison d'être of the *SCI*, and into the realms of sociometry, historiography, and science policy. Over the course of the last few decades, Garfield's innovative retrieval system begat a slew of unintended and unforeseen applications, and there is the prospect of much more to come (Cameron 1997).

PROSPECTS

Citation analysis is an important piece of the bibliometric research pie, one that will become even more central with the growth of the web. And for a very simple reason: the links (citations) provided routinely by authors in their reports and papers are a means of exposing the underlying sociocognitive structure of science. But links are also the defining feature of hypertext systems. As Larson (1996, 74) states: "[T]he notion of citation is fundamental both to the scholarly enterprise and to hypertext networks where it provides the primary mechanism for connection and traversal of the information space (or 'cyberspace')." The principles of citation indexing find their echo in the dynamically reticulated structure of the web, hence the proliferation of neologisms, such as cybermetrics, netometrics, webometrics, and influmetrics—the last of which was coined informally by Elisabeth Davenport to suggest diffuse and often imperceptible traces of scholarly influence—to capture the opportunities for measurement and evaluation afforded by the new environment. If citations can be tracked, counted, and weighted, then why not the links connecting websites? After all, citations and "sitations" are not merely similar phonetically, as Rousseau (1997) has noted—appropriately enough in *Cybermetrics*, a web-based journal. Highly linked sites are the web's equivalent of highly cited papers.

This thinking was axial to the work of the Clever Project (1999), whose team members explicitly acknowledged a methodological debt to Eugene Garfield and citation analysis. Specifically, they applied ISI's journal impact factor (see www.isinet.com/isi/hot/essays/7) to the evaluation of websites in order to identify "hubs" and "authorities," the web's analogues of citation "stars," nodes, or centroids in classical bibliometric studies. Others have come to the same conclusion; the search engine Google (see www .google.com/promote.html#search) uses link-based relevance weights to rank output. Going one step further, frequency of links, as with frequency of citation, can be construed as a proxy for social trust, as has recently been acknowledged both within and outside the literature of information science (Davenport and Cronin 2000; Schlossberg 1999).

On the web, a diverse array of sites, people, and objects can be linked to. This, potentially, facilitates the generation of multidimensional profiles of a scholar's or researcher's presence within (and across) particular communities of practice. As we (Cronin et al. 1998, 1326) posited some years ago, the "range of genres of invocation made possible by the Web should help give substance to modes of influence which have historically been backgrounded in narratives of science, or, at best, picked up, parenthetically, in biographies of the great and the good." In other words, more of those who are not particularly visible in terms of their publication or citation records (Meadows's [1974, 182] aforementioned "silent scientists") may finally receive due recognition for their sometimes unseen and unsung contributions (see Giles and Councill 2004). In addition to citations, we can track other forms of invocation, which, individually or conjointly, may provide more finely textured accounts of an individual's (or research team's) influence on current thinking, both locally and globally.

NEW PUBLISHING ENVIRONMENTS

The significance of the web from a bibliometric perspective goes well beyond enhanced opportunities for citation and link analysis. The web has challenged, and may revolutionize, many of the assumptions that have underpinned the established scholarly communication system. Radical proposals for open access and self-archiving have been put forward by a growing band of scholars, scientists, librarians, research administrators and others, some of whom were mentioned in chapter 4. Support for their pioneering ideas comes not only from significant sections of the grassroots scientific community (Markovitz 2000), but increasingly from the apex of the scientific establishment.

Harold Varmus, former director of the National Institutes of Health (NIH), the federal agency responsible for funding some $16 billion of biomedical research annually in the United States, was instrumental in conceiving and launching PubMed Central (née E-biomed). The aim, in short, is to create a web-based repository of biomedical research literature hosted by the NIH, to which global access will be provided free of charge (see www.pubmedcentral.nih.gov/). Commercial publishers are, naturally, less than thrilled by calls for free and open access to the literature of biomedicine and cognate fields. The creation of CrossRef by a consortium of some thirty commercial publishers—it uses the DOI (Digital Object Identifier) to link reference citations to full-text, online content held by different publishers (see www.crossref.org/)—may be construed, in part, as a response to unorthodox (and essentially public-sector-inspired) proposals for open access and institutional archiving. CrossRef, which describes itself as "the comprehensive source for linking journal articles," also provides further evidence of the fact that hypertext and citation indexing are a marriage made in heaven, though ISI may well view this pan-publisher initiative as a potential strategic threat to its de facto monopoly on the provision of large-scale, longitudinal citation data. Another pertinent development is the emergence of autonomous citation indexing as embodied in ResearchIndex, a value-adding system, developed at the NEC Research Institute in Princeton (now at Penn State's School of Information Sciences and Technology, sponsored by the NSF and Microsoft), that automatically extracts citations *and the context* in which the citations are made in the body of the citing paper (Lawrence, Giles, and Bollacker 1999). This allows the reader to see exactly how the citation relates to the surrounding text and to better understand both its instrumental and symbolic significance. Most recently, and for the first time, Giles and Councill (2004) have demonstrated that acknowledgment data can be automatically extracted from journal articles and analyzed along with citation data. Their pioneering study was carried out on CiteSeer using 335,000 computer science papers, more than 50 percent of which contained acknowledgments. Their automated study, an order of magnitude greater than any of the manual studies described in chapter 5, confirms the general validity and utility of acknowledgment analysis.

What is unclear in all of the ongoing experimentation and attendant speculation is the extent to which established peer review practices may be subverted by open peer review or, indeed, the extent to which different scientific communities and subfields may wish to adopt a tiered model of peer review, ranging from, say, full-bodied, double-blind refereeing to "peer review lite." It does, however, seem likely that the present monolithic publishing and peer review system will become rather more flexible and pluralistic in character as new publishing, posting, critiquing, and archiving behaviors establish themselves. Assuming that a more diverse publishing environment gradually emerges, bibliometricians will have a much broader array of objects and artifacts to feed into their accounts and analyses—both quantitative and qualitative—of scholars' communicative practices. This, inevitably, will turn the spotlight on a number of important issues relating to the provenance and durability (both intellectual and physical) of digital outputs: How reliable is a source? How credible is the producer/publisher? How persistent is the organ?

Traditionally, bibliometricians have dealt with an inherently stable environment—print-based publishing—and largely standard artifacts, outputs, and units of analysis (e.g., scholarly journals, peer-reviewed articles, citations). The new publishing environment, by way of contrast, promises heterogeneity of output and therewith greatly enhanced scope for, inter alia, quantifying multiple (and multimedia) outputs, tracing sociocognitive linkages, and harvesting online commentary/glosses on the works and ideas of professional scientists and scholars. Use statistics (whether of digital libraries, online journal collections, or other digital resources) have become a relatively direct by-product of web access and will likely feature in the evaluation of institution and individual performance (Giles and Councill 2004; Harnad and Brody 2004; Kurtz et al. 2004; Perneger, 2004). What, then, are the probable units of analysis associated with web-based publication and communication fora? What kinds of digital objects will actually be tracked, counted, weighted, and evaluated? In addition to the traditional journal article, we have electronic preprints, self-posted/self-archived documents, dynamically revised working papers, and multiple forms of "scholarly skywriting," to use one of Harnad's (1990, 342, 343) colorful metaphors. How, for instance, should we track, log, and assess such "sensitive measures of 'air time' and flight rate for new ideas and findings"? We also need to know how to identify, locate, and access these nontraditional (and, for now at least, often evanescent and ephemeral) outputs. And will common standards for identification, formatting, and labeling emerge, as well as for assuring the stability and persistence of less significant digital objects (e.g., online syllabi, courseware, threaded online chat, digital drafts, citations) over time?

In the present system, scholarly articles "belong to" particular journals, which, in the main, have stable existences and regular publication cycles. Persistent access to serials is provided by libraries with more or less coherent policies for collection development and management. In the New World order, bibliometricians and others will presumably want to know whether the objects of their scrutiny are part of a journal, host service, depository, discussion forum, website, or electronic archive. They will also need to have some shared sense of the quality, authority, and integrity of what is being mapped and measured. For instance, a scientific report hosted by the NIH's PubMed Central has, prima facie, higher credibility and cognitive authority than an unvetted opinion piece hosted on my personal server. All outputs are not created equal, and all citations are not equivalent. The warranting of claims in open publishing environments will most certainly be a more complex and nuanced matter than has been the case in the predigital communication system of science.

In a pluralistic and promiscuous publishing environment, the pedigree, provenance, and persistence of digital documents/objects, as well as their hosts, will be issues of consequence. Quality, in the present system, is signaled in a variety of ways. For example, a journal may be covered by major abstracting and indexing (A&I) services such as *Biosis*, *Medline*, or *Chemical Abstracts*, or the published/cited work may acknowledge funding from a research council, foundation, or other ostensibly reputable agency, which, typically, implies some degree of adjudication and also supports a reasonable presumption of quality. Our trust in texts is a function of the extent to which they have been screened or subjected to prepublication peer review in some form. The idea of open peer review on the web, or differential peer review, is gathering momentum, though concerns that the positive features (e.g., filtering) of the present peer review system—its acknowledged flaws notwithstanding (Chubin and Hackett 1990; Sense about Science 2004)—might be undermined, remain widespread.

As new technologies emerge and novel modalities of posting, hosting, publishing, filtering, and accessing information are developed, we can expect different disciplines and subfields to deploy quite different approaches and techniques. In addition, the rates of adoption of new models and behaviors will vary. The material practices of science differ greatly from one epistemic culture to another, and assumptions of common values and comparable metabolic rates are ill founded. As we have already seen, the sociocognitive structure of disciplines, and also their reward systems, differ greatly, which helps explain why, for instance, e-preprint exchange is so readily accepted within the rarefied world of high-energy physics but still regarded with suspicion by some other scientific communities. Universal approaches to communication, publication, filtering, and evaluation should certainly not be assumed; rather, pluralism, plasticity and adaptivity will be the hallmarks of the New World order.

VALIDITY CONCERNS

Critics of citation analysis (MacRoberts and MacRoberts 1989) have long challenged the assumption that citations can be used as valid indicators of quality, utility, or even impact, despite a compelling body of research findings to support the core contention (Baldi 1998; van Braam 1991). They have also identified reliability-related issues arising from the (perceived) selectivity and boundedness (e.g., in terms of language and geographic coverage) of ISI's coverage, though autonomous citation indexing may be a harbinger of unbounded and reliable citation indexing. But even if open hypertext systems—as envisaged by Cameron (1997)—allow us to get around the limitations of ISI's coverage, there is still the underlying problem of search engine reliability. Not only is there considerable variation between search engine retrieval performance but the same search engine will generate different output for the same search at different times (Bar-Ilan 2004; Rousseau 1998/1999; Snyder and Rosenbaum 1999). For this reason alone, extreme caution is advised in using web-derived, protoindicators of scholarly or communicative salience for evaluative purposes, most especially at the level of the individual.

Undoubtedly, construct validity issues will continue to surface, as new forms of web-based invocation are factored into bibliometric evaluations and sociometric narratives of scientific communication. Other measures may be less conventional but nonetheless deserving of consideration. Think of an individual whose ideas or theories constitute the cynosure of an extended listserv discussion within his or her peer community. Arguably, that in itself is an expression of perceived, if transient, utility or merit—whether the ideas are being lauded or rebutted. At present, there is no convenient or credible way of tracking or accommodating this kind of "scholarly skywriting" (Harnad 1990) in, for example, traditional faculty evaluations. It could be argued that the tenure process is already prolix and burdensome enough without adding yet more (and untried) variables to the evaluative mix, but that might just be myopic. The web is a fertile domain for scholarly expression and communication and emergent forms of recognition, acknowledgment, and approbation deserve to be documented and their social significance analyzed before being dismissed as invalid or supernumerary. Skeptics, as with the still-vocal doubting Thomases of classical citation analysis (Taubes 1993), will, quite rightly, want to know what various kinds of "sitations" and "hits" signify, and how we should go about distinguishing the substantive from the seriously trivial in open publishing environments.

Adler (1999a, 1999b) has coined the phrase "slashdot effect" to describe the spontaneous hit rate on a web server following the posting of a news story to a high-traffic site. He plotted the hit rate per minute for three stories published on *Slashdot*, *Linux Today*, and *Freshmeat* and found dramatic surge effects in each case. Convenient and appealing though such "splash factors" (Sayed-Brown 1999) might be, they could be of limited value. As the commercial market research community is only too well aware, there is a big difference between a consumer eye-balling a company's web-based advertising on a portal site and hyper-linking to that company's web page. Crude hits are much less useful measures than click-through rates in terms of assessing an advertisement's effectiveness—hence the growing interest in performance-based advertising (Gurley 1998). Analogously, hit rates to a scholar's home page should not be construed uncritically as surrogates of impact or quality but as, at most, preliminary indicators of possible impact (Thelwall, Vaughan, and Bjöneborn 2005).

As in academe, so in the marketplace. Something comparable to the phenomenon described by Adler seems to have happened following Victoria's Secret's advertising during the 1999 Super Bowl in the United States; hits to its website surged such that access degraded dramatically, a problem experienced, too, by Hotjobs.com, another first-time Super Bowl advertiser. Whether, and to what extent, these companies' profitability was boosted by their TV advertising (and the associated web spillover) is unclear. From an academic perspective, the question is whether the slashdot effect signifies a flash in the critical pan, or constitutes a leading and reliable indicator of scholarly impact.

Such surge-activating papers, postings, or provocations may sometimes prove to have little, if any, enduring academic, or other, merit. In web advertising terminology, it is the difference between click-through and look-to-buy metrics; the former record the number of visitors referred to a site by an online banner advertisement, while the latter compares banner impressions with subsequent sales transaction data (Schonberg et al. 2000). Furthermore, it has been found that website traffic estimates produced by firms such as Media Metrix and Nielsen NetRatings seem to significantly influence a company's stock market valuation (Thompson et al. 2000). The velocity and intensity of much web-based communication can easily create a false sense of what matters within a given community of scholars. The lessons are clear, and the challenge for bibliometricians in the new environment will be to understand that a gadarene rush does not necessarily equate with a posting's long-term scholarly or utilitarian significance. If anything, this kind of swarming behavior seems likely to intensify as recommender systems proliferate, and the numbers of pointers and links in circulation at any given moment grows (Terveen et al. 1997). But readers' time and attention are limited; the indiscriminate follower of others' recommendations may end up skimming and grazing in such a way that texts are rarely engaged with in meaningful fashion. In most cases, surge-activating papers will presumably exhibit very short, near-perpendicular life-cycle curves (so different from the typical life-cycle curve associated with a journal article's citation profile), as transient group interest races to the next hot item, creating a digital age equivalent of Andy Warhol's fifteen minutes of fame for web-present authors. As Waters (2004, 19) has observed in respect of humanities publishing, "Scholarly effect is measured in terms of depth, not width, of the reverberation the work sets off. Temporary celebrity is no guide at all."

CONCLUSIONS

"Bibliographic citation," as Harnad and Carr (2000, 629) have observed, "is the mother of all hyperlinks." With the creation of the web, its time has unquestionably come. The application, and extension, of existing bibliometric techniques to open publishing environments should, ultimately, result in the development of a portfolio of metrics for better capturing the totality of ways in which cognitive influence is exercised and exhibited within and across specialty groups. One of the limitations of the ISI's databases is that they are restricted to a relatively small, albeit high-quality, subset of the universe of scholarly journals. This means that citations to a scholar's work that appear in journals (not to mention monographs) excluded from ISI's coverage are, to all intents and purposes, lost (Cronin, Snyder, and Atkins 1997). Additionally, journals not included in ISI's coverage are less visible than they might otherwise be, an understandably sore point for some peripheral nations, especially since many of the papers cited by the journals covered by ISI's indexes are actually published in journals that lie beyond the citational pale (Maricic 1999). However, as the developers of ResearchIndex have shown, a new generation of citation (and, indeed, acknowledgment) analysis tools is emerging, one that is not subject to the kinds of commercial and technical constraints that have necessarily inhibited the development of ISI's citation indexing products. Also, at the time of writing, the Open Society Institute has convened a working group to develop an Open Access Citation Index, though no concrete developments have been reported as yet.

The hypertextual character of the web means that the principles of citation indexing can be applied much more widely than at present. Web-based retrieval systems will allow us to go beyond traditional citations and track acknowledgments, diffuse contributions, and other input measures, a fact grasped by, among others, the Wellcome Trust (Lewison 2000). Broadly conceived, web-based citation indexing and analysis should ensure that the sometimes overlooked inputs and influences of technicians, mentors, trusted assessors, and sundry collaborators (Mullins 1973) could more easily be factored into the recognition and assessment calculus. That said, great care will be needed to ensure that the symbolic significance of such signs—these "pellets of . . . peer recognition," to reprise Merton's (2000, 438) euphonious phrase—is not misinterpreted and subsequently misused.

Citation context analysis and deep linking (pointing to a precise spot on a website) will allow us to explore the contexts in which invocations occur and thereby develop a more reliable sense of why an individual warrants mention, or how an individual's work and ideas are perceived and received by his or her peer community. On the web, scholars do more than publish, or post, their working papers and finished articles; they spray ideas, debate issues, and launch trial balloons, in ways that occasionally deviate from traditional practice. And they recommend their own work, and the work of selected others, to their peers. This, of course, is what we do (most of the time) when we cite others. Citation indexes may thus be viewed as a prototypical species of recommender system (Terveen et al. 1997). In the near future, web-based, bibliometric spectroscopy will allow us to map with greater fidelity the matrix of informal endorsements and recommendations which lie behind the formal rating systems of science—"social indicator mining," if you will. Concrete applications might include cocitation analysis of hyperlinks to reveal the organization of online communities and their knowledge bases (Pitkov and Pirolli 1997). It is equally conceivable that next-generation search engines and bibliometric tools will allow us to observe "science in action" (Bossy 1995), in the process enabling us to detect early signs of emerging trends and also develop a better sense of those scientists and researchers who have what might be termed "street cred," as opposed to the established coin of the realm, eminence.

Much remains unclear about the kinds of emergent practices and protoindicators introduced in this chapter and their longer-term import within the academic evaluation context but, here, surely is an issue ripe for systematic scrutiny, as Harnad (2004) acknowledges: "Citation counts do not measure quality directly, but they are correlated with it. So are download counts, and no doubt other digitometric measures that are under development, and that will be derived from a growing OA corpus." Further evidence of the effect of open access on citation impact has been provided by Brody et al. (2004), Harnad and Brody (2004), and Perneger (2004). As the competition for attention intensifies, the tools for tracking eyeballs and measuring impact are becoming progressively more sophisticated. We can expect to see novel indicators of peer recognition and attention emerge and take their place alongside traditional bibliometric measures, such as publications counts, citation scores and impacts factors, in the very near future.

REFERENCES

Adler, S. (1999a). The Slashdot Effect: an analysis of three Internet publications. Available at http://ssadler.phy.bnl.gov/adler/SDE/SlashDotEffect.htm.

Adler, S. (1999b). Addendum to the Slashdot Effect Internet paper. 1999. Available at http://ssadler.phy.bnl.gov/adler/SDE/SlashDotEffectAddendum.html.

Baldi, S. (1998). Normative versus social constructivist processes in the allocation of citations: A network analytic model. *American Sociological Review*, 63, 829–846.

Bar-Ilan, J. (2004). The use of Web search engines in information science research. In: Cronin, B. (Ed.). *Annual Review of Information Science and Technology*, 38. Medford, NJ: Information Today, 231–288.

Becher, T. (1989). *Academic Tribes and Territories: Intellectual Enquiry and the Culture of disciplines*. Milton Keynes: Oxford University Press.

Ben-Ari, E. (1987). On acknowledgements in ethnographies. *Journal of Anthropological Research*, 43(1), 63–84.

Berger, J., Wagner, D. G., and Zelditch, M. (1989). Theory growth, social processes, and metatheory. In: Turner, J. H. (Ed.). *Theory Building in Sociology: Assessing Theoretical Cumulation*. Sage: London, 19–42.

Bossy, M. J. (1995). The last of the litter: "Netometrics." *Solaris*, 2. Available at http://www.info.unicaen.fr/bnum/jelec/Solaris/d02/2bossy.html.

Bourdieu, P. (1990). *Homo Academicus*. Cambridge: Polity Press.

Bourdieu, P. (1991). *Language and Symbolic Power*. (J. B. Thompson, Ed.) Cambridge, MA: Harvard University Press.

Brodkey, L. (1987). *Academic Writing as Social Practice*. Philadelphia: Temple University Press.

Brody, T., Stamerjohanns, H., Harnad, S., Gingras, Y., Vallieres, F., and Oppenheim, C. (2004). The effect of open access on citation impact. Paper presented at *National Policies on Open Access (OA) Provision for University Research Output: an International Meeting*. Southampton University, 19th February 2004. Available at http://opcit.eprints.org/feb19prog.html.

Cameron, R. D. (1997). A universal citation database as a catalyst for reform in scholarly communication. Available at http://elib.cs.sfu.ca/project/papers/citebase/citebase.html.

Chubin, D. E. (1975). Trusted assessorship in science: a relation in need of data. *Social Studies of Science*, 5, 362–368.

Chubin, D. E. and Hackett, E. J. (1990). *Peerless Science: Peer Review and U.S. Science Policy*. Albany: State University of New York Press.

Clever Project. (1999). Hypersearching the Web. *Scientific American*, 54–60.

Cronin, B. (1981). The need for a theory of citing. *Journal of Documentation*, 37(1), 16–24.

Cronin, B. (1984). *The Citation Process: The Role and Significance of Citations in Scientific Communication*. London: Taylor Graham.

Cronin, B. (1994). Tiered citation and measures of document similarity. *Journal of the American Society for Information Science*, 45(7), 537–538.

Cronin, B. (1995). *The Scholar's Courtesy: The Role of Acknowledgement in the Primary Communication Process*. London: Taylor Graham.

Cronin, B. (1996). Rates of return to citation. *Journal of Documentation*, 52(2), 188–197.

Cronin, B., Snyder, H., and Atkins, H. (1997). Comparative citation rankings of authors in monographic and journal literature: a study of sociology. *Journal of Documentation*, 53(3), 263–273.

Cronin, B., Snyder, H. W., Rosenbaum, H., Martinson, A., and Callahan, E. (1998). Invoked on the Web. *Journal of the American Society for Information Science*, 49(14), 1319–1328.

Czarniawska, B. (1997). *Narrating the Organization: Dramas of Institutional Identity*. Chicago: University of Chicago Press.

Davenport, E. and Cronin, B. (2000). The citation network as a prototype for representing trust in virtual environments. In: Cronin, B. and Atkins, H. B. (Eds.). *The Web of Knowledge: A Festschrift in Honor of Eugene Garfield*. Medford, NJ: Information Today, 517–534.

Garfield, E. (1955). Citation indexes for science: A new dimension in documentation through association of ideas. *Science*, 122, 108–111.

Giddens, A. (1984). *The Constitution of Society*. Berkeley, CA: University of California Press.

Giles, C. L. and Councill, I. G. (2004, Dec. 21). Who gets acknowledged? Measuring scientific contributions through automatic acknowledgement indexing. Proceedings of the National Academy of Sciences 101(51), 17599–17604.

Grafton, A. (1997). *The Footnote: A Curious History*. Cambridge, MA: Harvard University Press.

Gurley, W. (1998, November 9). How the Web will warp advertising. *Fortune*, 119–120.

Hagstrom, W. O. (1982). Gift giving as an organizing principle in science. In: Barnes, B. and Edge, D. (Eds.). *Science in Context: Readings in the Sociology of Science*. Cambridge, MA: MIT Press, 21–34.

Harnad, S. (1990) Scholarly skywriting and the prepublication continuum of scientific inquiry. *Psychological Science*, 1, 342–343.

Harnad, S. (2004, June 5). Impact factor, open access & other statistics-based quality. Posted to SIGMETRICS@LISTSERV.UTK.EDU

Harnad, S. and Brody, T. (2004). Comparing the impact of open access (OA) vs. non-OA articles in the same journals. *D-Lib Magazine*, 10(6). Available at www.dlib.org/dlib/june04/harnad/06harnad.html.

Harnad, S. and Carr, L. (2000). Integrating, navigating, and analysing open Eprint archives through open citation linking (the OpCit project). *Current Science*, 79(5), 629–638.

Hemlin, S. (1996). Research on research evaluation. *Social Epistemology*, 10(2), 209–250.

Hicks, D. and Potter, K. (1991). Sociology of scientific knowledge: A reflexive citation analysis *or* science disciplines and disciplining science. *Social Studies of Science*, 21, 459–501.

Hjørland, B. (1997). *Information Seeking and Subject Representation: An Activity-Theoretical Approach to Information Science*. Westport, CT: Greenwood.

Jevons, F. R. (1973). *Science Observed: Science as a Social and Intellectual Activity*. London: Allen & Unwin.

Kurtz, M. J., Eichorn, G., Accomazzi, A., Grant, C., Demleitner, M., Murray, S. S., Martimbeau, N., and Elwell, B. (2004). The bibliometric properties of article readership information. *Journal of the American Society for Information Science and Technology* (in press).

Latour, B. and Woolgar, S. (1982). The cycle of credibility. In: Barnes, B. and Edge, D. (Eds.). *Science in Context: Readings in the Sociology of Science*. Cambridge, MA: MIT Press, 35–43.

Larson, R. R. (1996). Bibliometrics of the World Wide Web: An exploratory analysis of the intellectual structure of cyberspace. In: Hardin, S. (Ed.). *Global Complexity: Information, Chaos and Control. Proceedings of the 59th ASIS Annual Meeting. Baltimore, MD. October 21–24, 1996*. Medford, NJ: Information Today Inc., 71–78.

Lawrence, S., Giles, C. L., and Bollacker, K. (1999). Digital libraries and autonomous citation indexing. *IEEE Computer*, 32(6), 67–71.

Leydesdorff, L. (1998). Theories of citation? *Scientometrics* 43(1), 5–25.

Lewison, G. (2000). Citations as a means to evaluate biomedical research. In: Cronin, B. and Atkins, H. B. (Eds.). *The Web of Knowledge: A Festschrift in Honor of Eugene Garfield*. Medford, NJ: Information Today, 361–372.

Macintosh, N. B. and Scapens, R. W. (1990). Structuration theory in management accounting. *Accounting, Organizations, and Society*, 15(5), 455–477.

MacRoberts, M. (1997). Letter to the editor. *Journal of the American Society for Information Science*, 48(10), 963.

MacRoberts, M. H. and MacRoberts, B. R. (1989). Problems of citation analysis: a critical review. *Journal of the American Society for Information Science*, 40(5), 342–349.

Maricic, S. (1999, January 28). Mainstream-periphery science communication. *Nature*. Available at http://helix.nature.com/wcs/e01.html.

Markovitz, B. P. (2000). Biomedicine's electronic publishing paradigm shift: Copyright policy and PubMed Central. *Journal of the American Medical Informatics Association*, 7, 222–229.

McCain, K. W. (1991). Communication, competition, and secrecy: The production and dissemination of research-related information in genetics. *Science, Technology, & Human Values*, 16(4), 491–516.

Meadows, A. J. (1974). *Communication in Science*. London: Butterworths.

Merton, R. K. (2000). On the Garfield input to the sociology of science: A retrospective collage. In: Cronin, B. and Atkins, H. B. (Eds.). *The Web of Knowledge: A Festschrift in Honor of Eugene Garfield*. Medford, NJ: Information Today, 435–448.

Mitroff, I. I . (1974). Norms and counter norms in a select group of Apollo moon scientists: a case study of the ambivalence of scientists. *American Sociological Review*, 39, 579–595.

Mullins, N. C. (1973). *Theories and Theory Groups in Contemporary American Sociology*. London: Harper & Row.

Pitkow, J. and Pirolli, R. (1997). Life, death, and lawfulness on the electronic frontier. *Proceedings CHI 97, Conference on Human Factors in Computing Systems, 22–27 March 1997, Atlanta, Georgia. USA*. Available at www.acm .org/sigchi/chi97/proceedings/paper/jp-www.htm.

Rosenbaum, H. (1996). *Managers and Information in Organizations: Towards a Structurational Concept of the Information Use Environment of Managers*. Syracuse, NY: Graduate School Syracuse University, PhD thesis.

Rousseau, R. (1997). Sitations: an exploratory study. *Cybermetrics* 1(1). Available at www.cindoc.csic.es/cybermetrics/vol1iss1.html.

Rousseau, R. (1998/99). Daily time series of common single word searches in AltaVista and Northern Light. *Cybermetrics*, 2/3(1). Available at www .cindoc.csic.es/cybermetrics/articles/v2i1p2.html.

Sayed-Brown, C. (1999, February 5). E-hits and tenure? Open Lib/Info Sci Education Forum. JESSE@UTKVM1.UTK.EDU.

Schlossberg, E. (1999). A question of trust. *Brill's Content*, 2(2), 68–70.

Schonberg, E., Cofino, T., Hoch, R., Podlaseck, M., and Spraragen, S. L. (2000). Measuring success. *Communications of the ACM*, 43(8), 53–57.

Sense About Science. (2004). *Peer Review and the Acceptance of New Scientific Ideas. Discussion Paper from a Working Party on Equipping the Public with an Understanding of Peer Review*. London: Sense About Science.

Shadish, W. R. (1989). Perceptions and evaluation of quality in science. In: Gholson, B., Shadish, W. R., Neimeyer, R. A., and Houts, A. C. (Eds.). *Psychology of Science. Contributions to Metascience*. Cambridge: Cambridge University Press, 383–426.

Shapin, S. (1994). *A Social History of Truth*. Chicago, IL: University of Chicago Press.

Silverstone, R., Hirsch, E., and Morley, D. (1992). Information and communication technologies and the moral economy of the household. In: Silverstone, R. and Hirsch, E. (Eds.). *Consuming Technologies: Media and Information in Domestic Spaces*. London: Routledge, 15–31.

Small, H. (1978). Cited documents as concept symbols. *Social Studies of Science,* 7, 142–147.

Snyder, H. and Rosenbaum, H. (1999). Can search engines be used as tools for web-link analysis? A critical view. *Journal of Documentation*, 55(4), 375–384.

Suchman, L. (1996). Supporting articulation work. In: Kling, R. (Ed.). *Computerization and Controversy: Value Conflicts and Social Choices*. New York: Academic Press, 407–423.

Swanson, D. R. (1989). A second example of mutually isolated medical literatures related by implicit, unnoticed connections. *Journal of the American Society for Information Science*, 40(6), 432–435.

Taubes, G. (1993). Measure for measure in science. *Science*, 260, 884–886.

Terveen, L., Hill, W., Amento, B., McDonald, D., and Creter, J. (1997). PHOAKS: A system for sharing recommendations. *Communications of the ACM*, 40(3), 59–62.

Thelwall, M., Vaughan, L. and Björneborn, L. (2005). Webometrics. In: Cronin, B. (Ed.). *Annual Review of Information Science and Technology*, 39. Medford, NJ: Information Today, 81–135.

Thompson, M. J., Lawrence, S., Lake, D., and Mowrey, M. A. (2000, July 24). Trading on ratings. *The Industry Standard*, 171, 173–175, 178–179, 182–183, 186, 188, 190, 193.

Turner, J. H. (1989). Introduction: can sociology be a cumulative science? In: Turner, J. H. (Ed.). *Theory Building in Sociology: Assessing Theoretical Cumulation*. London: Sage, 8–18.

Turner, S. (1994). The origins of 'mainstream sociology' and other issues in the history of American sociology. *Social Epistemology*, 8(1), 41–67.

van Braam, R. R. (1991). *Mapping of Science: Foci of Intellectual Interest in Scientific Literature*. University of Leiden: DSWO Press.

van Raan, A. F. J. (2000). The Pandora's Box of citation analysis: measuring scientific excellence, the last evil? In: Cronin, B. and Atkins, H. B. (Eds.). *The Web of Knowledge: A Festschrift in Honor of Eugene Garfield*. Medford, NJ: Information Today, 301–319.

Vakkari, P. (1997). Information seeking in context: a challenging metatheory. In: Vakkari, P., Savolainen, R., and Dervin, B. (Eds.). *Information Seeking in Context*. London: Taylor Graham, 451–464.

Vakkari. P. and Kuokkanen, M. (1997). Theory growth in information science: applications of the theory of science to a theory of information seeking. *Journal of Documentation*, 53(5), 497–519.

Vickery, B. (1997). Metatheory and information science. *Journal of Documentation*, 53(5), 457–476.

Warner, J. (1990). Semiotics, information science, documents and computers. *Journal of Documentation*, 46(1), 16–32.

Waters, L. (2004). *Enemies of Promise: Publishing, Perishing, and the Eclipse of Scholarship*. Chicago: Prickly Paradigm Press.

Weick, K. E. and Roberts, K. (1993). Collective mind in organizations: Heedful interrelating on flightdecks. *Administrative Science Quarterly*, 38, 357–381.

Weinberg, B. H. (1997). The earliest Hebrew citation indexes. *Journal of the American Society for Information Science*, 48(4), 318–330.

White, H. D. (1990). Author co-citation analysis: overview and defense. In: Borgman, C. L. (Ed.). *Scholarly Communication and Bibliometrics*. Newbury Park, CA: Sage, 84–106.

White, H. D. (2004). Citation analysis and discourse analysis revisited. *Applied Linguistics*, 25(1), 89–116.

Wouters, P. (1999). *The Citation Culture*. University of Amsterdam, PhD thesis.

Chapter Nine

Scientometric Spectroscopy

A colleague suggested that *The Hands of Science* would be a better title for this book, the dominant themes being collaboration and teamwork. She has a point; but I am loath to forego the scribal allusion, clever though the plural form is. This much is clear, albeit an oversimplification: the doing of science is less about the mechanics and stylistics of academic writing than about the conceptualization, design, and conduct of experiments; the gathering and evaluation of evidence; and the progressive refinement of theories, models, and assumptions. The papers that result from laboratory tests, clinical trials, and fieldwork are in one sense afterthoughts: important elements of the whole that must conform to prevailing genres (while dutifully allocating credit) but not necessarily, in themselves, the most crucial or labor-intensive aspect of the research process. In passing, it bears remarking that in some epistemic cultures peer recognition and reward are not contingent upon formal publication. On occasion, insider knowledge may matter more than inscription. Gusterson (2003, 293) has shown how in the localized community of nuclear weapons designers, credit is "established and circulated as much by word of mouth . . . as through the written documentation of individual contributions and achievements."

In physics, as we have seen, many hands make heavy work lighter. Hundreds of individuals may be involved in multiyear experiments and in due course all of their names will appear on the masthead, but we know perfectly well that very few of those headlined have been materially involved in the writing or editing of the publication of record. They are authors in name only. So, too, in the biomedical research community, though here the trend to explicitly record researchers' differential contributions helps establish exactly which hands did what. It is no exaggeration to say that authorship, as traditionally construed, is something of an anachronism in contemporary science, a trope in need of retreading.

But we must not assume that what holds for certain highly specialized scientific fields applies to all branches of research and scholarship. As Biagioli and Galison (2003, 1) have observed, the "relationship between the author's name and the epistemological status of claims vary across different disciplines." This little book is indubitably the handiwork of a sole author. Its structure, biases, style, and tone are very much mine and, moreover, some of the ideas contained herein only crystallized in the act of writing. Of course, the original empirical findings that gave rise to the monograph were, as I have pointed out in the acknowledgments, the result of *formal* collaborations (that is to say, I at times had coworkers and coauthors). The foundational papers arising from those modest collaborations in turn recognized a number of individuals whose *informal* contributions added value in one way or another to the final products. Ultimately, however, *The Hand of Science*, viewed primarily as a literary artifact, is the work of one individual (one hand); viewed primarily as a cognitive artifact, it is the work of a distributed coalition (many hands), albeit, for the record, a coalition in which effort and originality were quite differentially contributed.

In recent years, Science and Technology Studies (STS) has brought an impressive array of theoretical and methodological perspectives to bear on role of science and technology in society, including the processes whereby knowledge is constructed, validated, shared, and utilized (van House 2004). Scientific knowledge, so it is argued, "both embeds and is embedded in social practices, identities, norms, conventions, discourses, instruments and institutions—in short, in all the building blocks of what we term the social" (Jasanoff 2004a, 3). Such claims, now commonplace, are borne out by the evidence presented in this book. We have seen how the norms and values that constitute different epistemic cultures shape, and in turn are shaped by, the technologies scholars use to publish their work and to facilitate communication with both their peers and the public at large. We have also seen how social networks and personal/professional relationships are reflected in authors' citation (and acknowledgment) images and identities. And we have seen how the use made of information and communication technologies is as much a function of social values and local practices as the features and functionalities of the tool sets themselves. Gradually, deterministic accounts of information technology and its "impacts'" are being replaced by more culturally sensitive, sociotechnical analyses that strive to capture the particularities of local practice.

Kling, McKim, and King (2003), in their study of scholarly communication practices in the fields of high-energy physics, molecular biology, and information systems, came to consider information technologies as "socio-technical interaction networks," a more nuanced and complex perspective than that traditionally favored by most information systems analysts. The idea of coproduction (of the social and the technical, that is), implicit in the title of Kling et al.'s paper, is also favored by Jasanoff (2004a, 5), who, in her edited volume *States of Knowledge: The Co-Production of Science and Social Order*, demonstrates the growing appeal and analytic utility of what she variously labels "the co-productionist framework" or "co-productionist idiom" in helping us understand events and phenomena in the world. Those, she argues, who would provide accounts of science-in-action need to identify or develop suitable methods to capture the "constant intertwining [note the Klingian echo here] of the cognitive, the material, the social and the normative" (Jasanoff 2004a, 6). Or, to put it otherwise and most succinctly: "Co-production . . . sacrifices simplicity for richness and linearity for deeper contextualization" (Jasanoff 2004b, 280).

The scripts that scholars routinely produce in the course of their work do not emerge from a social vacuum; rather, they are the products of authors' continual interactions with professional colleagues, students, and others. Sometimes, the research collaboration is large and structurally complex; at other times, it is craft-level and easily manageable. And, as we have seen, not all collaboration is formal in nature. Admittedly, coauthorship data can tell us a great deal about social and interinstitutional linkages, while citation maps can illuminate otherwise dimly grasped webs of intellectual ties between individuals, groups, disciplines, and nations, but there remains a wealth of complementary sociometric data buried (and waiting to be exploited) in acknowledgments—a heretofore largely ignored facet of the scientific journal article, one that testifies to the (social) situatedness of scientists and scholars, no matter what their field or discipline. In their automated extraction and analysis of acknowledgments in the online literature of computer science, Giles and Councill (2004, 7) found that "a surprising degree of intellectual debt to individuals [is] documented through the mechanism of acknowledgment." In fact, their findings are not really "surprising" (Cronin 1995), but it is heartening to see that this particular paratextual vein can now be quarried with relative ease and on a much larger scale than has been possible heretofore. Their study is a harbinger of things to come.

With the prodigious growth in digital libraries, institutional repositories and open archives, we can expect to see a flurry of interest in undertaking new kinds of (socio-) cybermetric studies. There will soon be a critical mass of web-based digital objects and usage statistics on which to model scholars' communication behaviors—publishing, posting, blogging, scanning, reading, downloading, glossing, linking, citing, recommending, acknowledging—and with which to track their scholarly influence and impact, broadly conceived and broadly felt (e.g., Feitelson and Yovel 2004; Giles and Councill 2004; Harnad et al. 2003; Kurtz et al. 2004; Lawrence 2001; Thelwall and Harries 2003). We are moving, if I may tweak van Raan's metaphor, into the age of scientometric spectroscopy.

REFERENCES

Biagioli, M. and Galison, P. (2003). Introduction. In: Biagioli, M. and Galison, P. (Eds.). *Scientific Authorship: Credit and Intellectual Property in Science*. New York: Routledge, 1–9.

Cronin, B. (1995). *The Scholar's Courtesy: The Role of Acknowledgement in the Primary Communication Process*. London: Taylor Graham.

Feitelson, D. G. and Yovel, U. (2004). Predictive ranking of computer scientists using CiteSeer data. *Journal of Documentation*, 66(1), 44–61.

Giles, C. L. and Councill, I. G. (2004, Dec. 21). Who gets acknowledged? Measuring scientific contributions through automatic acknowledgement indexing. Proceedings of the National Academy of Sciences 101(51), 17599–17604.

Gusterson, H. (2003). The death of authors of death: Prestige and creativity among nuclear weapons scientists. In: Biagioli, M. and Galison, P. (Eds.). *Scientific Authorship: Credit and Intellectual Property in Science*. New York: Routledge, 282–307.

Harnad, S., Carr, L., Brody, T., and Oppenheim, C. (2003). Mandated online RAE CVs linked to university eprint archives: Enhancing U.K. research impact and assessment. *Ariadne*, 35. Available at: http://www.ariadne.ac.uk/issue35/harnad/intro.htm

Jasanoff, S. (2004a). The idiom of co-production. In: Jasanoff, S. (Ed.). *States of Knowledge: The Co-Production of Science and Social Order*. New York: Routledge, 1–12.

Jasanoff, S. (2004b). Afterword. In: Jasanoff, S. (Ed.). *States of Knowledge: The Co-Production of Science and Social Order*. New York: Routledge, 274–282.

Kling, R., McKim, G. and King, A. (2003). A bit more to IT: Scholarly communication forums as socio-technical interaction networks. *Journal of the American Society for Information Science and Technology*, 54(1), 47–67.

Kurtz, M. J., Eichorn, G., Accomazzi, A., Grant, C., Demleitner, M; Murray, S. S., Martimbeau, N., and Elwell, B. (2004). The bibliometric properties of article readership information. *Journal of the American Society for Information Science and Technology* (in press).

Lawrence, S. (2001). Online or invisible? *Nature*, 411(6837), 521.

Thelwall, M. and Harries, G. (2003) The connection between the research of a university and counts of links to its Web pages: An investigation based upon a classification of the relationships of pages to the research of the host university. *Journal of the American Society for Information Science and Technology*, 54(7), 594–602.

van House, N. A. (2004). Science and Technology Studies and Information studies. In: Cronin, B. (Ed.). *Annual Review of Information Science and Technology*. 38, Medford, NJ: Information Today, 3–86.

van Raan, A. F. J. (2000). The Pandora's Box of citation analysis: Measuring scientific excellence—the last evil? In: Cronin, B. and Atkins, H. B. (Eds.). *The Web of Knowledge: A Festschrift in Honor of Eugene Garfield*. Medford, NJ. Information Today, 301–319.

Index

Woolgar, S., 1, 11, 71, 170
work/object involved, 148
Worm Community System, 62
Wouters, P., 143, 145, 146,
    148–49
writing: as instance of distributed
    cognition, 74, 86; linked to
    authorship, 58, 63; as social act,
    108–10, 154

Yank, V., 51, 59, 60, 109
Yates, S. J., 155

Zhou, X., 119
Zhu, J., 127
Ziman, J., 86
Zipf, George Kingsley, 81
Zuccala, A., 104–5
Zuckerman, H., 98